Ivy Panagassi
Andréia M. Pancher

Interpretive trail in a conservation unit

Ivy Panagassi
Andréia M. Pancher

Interpretive trail in a conservation unit

Environmental education tool

ScienciaScripts

Imprint

Cover image: www.ingimage.com

This book is a translation from the original published under ISBN 978-3-330-75952-7.

Publisher:
Sciencia Scripts
is a trademark of
Dodo Books Indian Ocean Ltd. and OmniScriptum S.R.L publishing group

120 High Road, East Finchley, London, N2 9ED, United Kingdom
Str. Armeneasca 28/1, office 1, Chisinau MD-2012, Republic of Moldova, Europe
Printed at: see last page
ISBN: 978-620-8-31017-2

SUMMARY

I dedicate this work to my family.

ACKNOWLEDGMENTS

Thank you for the support and love of my family (Val, My, Gui, André, Nilson, Grandma Marlene, Grandpa Bruxo, Grandpa Joao and Grandma Ivone (in memoriam), Aunt Vi, Uncle Màrcio, Uncle Gilson, Uncle Denilson, Uncle Lei, Adrienne, Aunt Cris, Caio, Lola, Bia, Sofia, Enrico, Helena, Cecilia, Marina, Penèlope, Stafanie and Doroty (in memoriam) Chico, Pietro, Goiaba, Diego). Special thanks to the people who raised me and from whom I learned the values and principles that guide my life, my mothers Val, My and my double mother Grandma Marlene; and to my father Nilson. Special thanks also to my brother Gui, an example of a human being, and to my partner André for his affection, patience and dedication. To all the plants that inhabit my home, which give me great moments of joy and teach me to have affection and love for all living beings.

I would like to thank the Management Department team (Timoni, Fabi, Zezao, Keila, Nathalie, Monique, Andreia, Marcelo, Karina, Seu Augustinho, Làzaro) for their help in carrying out the research. To the employees of the Forestry Foundation who make up FEENA's management team. To the environmental monitors from the BK company (Màrcia, Mococa, Pi). To the staff of the Rio Claro Tourism Office.

To my advisor Andréia Medinilha Pancher for her guidance and contribution.

and if we're left with nothing but disappointment,
collapse, lapse, kidnapping, corruption?
and more disgrace, more degradation?
concentration, maldistribution?
...
and it's all ruin, all under construction
and the vast jungle, only devastation
so our contribution
is it not just song, consolation?
is there no other solution?
...
but if we have plans, and they are
the end of hunger and defamation
why not put them into action straight away?
so now is the inauguration
of our new civilization
as unique as our own
and be beautiful, free, luminous
our dreams of nationhood.
...
we have to do it precisely
the distinction between project and dream
to dream at last without illusion
the luminous dream of reason

Echoes of ao - Lenin

SUMMARY

Within the current context of environmental degradation, caused mainly by the unsustainable pattern of production and consumption (AGENDA 21, 1992), Conservation Units (CUs) are internationally considered to be one of the most successful measures for conserving biodiversity, according to the Union for the Conservation of Nature (IUCN). The city of Rio Claro (Sao Paulo, Brazil) has an important Conservation Unit in both environmental and historical-cultural aspects, the 'Edmundo Navarro de Andrade' State Forest (FEENA).In addition to conservation measures, there is a need to develop environmental education activities in Conservation Units that encourage community participation and appreciation in the maintenance and protection of the unit, as well as providing moments of reflection that can trigger perceptual, behavioral, attitudinal and evaluative changes in the relationship between human beings and their surroundings. This work proposes the creation of an interpretive trail in FEENA's public use area to contribute to the Unit's Management Plan as an environmental education tool. In order to plan the interpretive trail, a bibliographical and cartographic review was carried out on the subject, questionnaires were applied in order to find out about the visiting public and the characteristics of public visitation, as well as visitors' preferences in terms of receiving information about the area and fieldwork at the trail site. The trail was designed with environmental and social aspects in mind. To this end, a trail suspended from the ground is proposed, with the aim of impacting the area as little as possible and including people with reduced mobility and the visually impaired in the context of the UCs.

Keywords: Environmental Education; Interpretive Trail; Conservation Units;

1 INTRODUCTION

The model of exploitation of natural, vegetation and water resources established from the 1950s onwards has allowed every possible type of environmental aggression, proving harmful to the environment to this day, destroying large parts of our forests, riparian forests, water resources and even causing the loss of our fauna (verbal information) .[1]

The alarming urban growth of Brazilian cities has drastically reduced opportunities for interaction with the environment, particularly for low-income populations. The remaining natural areas available in metropolises are generally in two different contexts: either they are very small, or they are degraded to such an extent that they require multidisciplinary programs to restore them.

The speed and irreversibility of the processes of exploitation of natural resources require a variety of measures to conserve the environment. One of the logical ways of implementing these measures is through the creation and management of Conservation Units (CUs), which are legally protected spaces with relevant natural characteristics.

The city of Rio Claro, located in the interior of the state of São Paulo, has an important Conservation Unit, the 'Edmundo Navarro de Andrade' State Forest (FEENA), which stands out for its environmental and historical-cultural aspects.

In view of the growth of the urban area of the city of Rio Claro and the environmental degradation observed in the state of São Paulo, FEENA plays a vital role in stabilizing and minimizing socio-environmental impacts and the quality of life of the residents of the city and region. The main functions of this forest are:

- Climate stabilization by preventing the advance of heat islands;
- Atmospheric recovery by filtering polluted air;
- Water supply and protection of water sources by preserving three important rivers responsible for supplying water to 20% of Rio Claro's population;
- Protection of biodiversity, as it is home to different species of native plants and wild animals, despite being the result of reforestation (verbal information) .[2]

FEENA has an important history in the region and is closely linked to the emergence of the city of Rio Claro, when Antônio Paes de Barros, the first Baron of Piracicaba, lived on the land where the UC now stands. It is also considered a legacy for the forestry studies left by Edmundo Navarro de

---

1 Information provided by Timoni during an internship at the Rio Claro City Hall's Planning, Development and Environment Department in 2014.
2 Information provided by Timoni during an internship at the Rio Claro City Council's Planning, Development and Environment Department (SEPLADEMA) in 2014.

Andrade and a landmark for reforestation aimed at producing raw materials for the railroad.

As well as protecting biodiversity, the landscape and natural resources, the PAs play a fundamental role in ecological and historical-cultural tourism in the region, protecting natural and historical heritage and proving vital in promoting environmental education.

Interpretive trails located in natural environments, such as UCs, play a pedagogical role that is an important tool for the development of Environmental Education. They allow visitors to come into direct contact with nature, sharpening their capacity for observation and logical reasoning. It is capable of combining an educational process with recreational practice in order to instigate and promote reflection on the importance of conserving natural assets, linked to the history of territorial formation and the identity of the local population. These reflections could trigger perceptual, behavioral, attitudinal and evaluative changes about oneself and the environment in which one lives.

In order to effectively fulfill its objectives, the interpretive trail must guarantee nature conservation and, at the same time, ensure awareness, learning and perception of the environment, all in an attractive and captivating way.

In order to contribute to the Management Plan of the Conservation Unit in question, with regard to the area of environmental education and interpretation and public use, this work proposes the creation of an interpretive trail in the Public Use Area of the Forest.

In order to plan the interpretive trail, it was necessary to carry out a bibliographical review on the subject, as well as to deepen our knowledge of FEENA's Management Plan, in order to comply with the rules and guidelines proposed for it. In order to get to know the visiting public and the characteristics of public visits, as well as visitors' preferences for receiving information about the study area, questionnaires were administered to visitors as a methodological resource. This step was also essential for good planning, as it allowed us to get to know the target audience of the proposal.

Based on knowledge of the laws relating to Conservation Units, the physical, biological and socio-cultural aspects of the site, carried out through fieldwork, a bibliographical and cartographic survey of the area, together with the characteristics of the visiting public and the Management Plan consulted, it was possible to determine precisely the location and route of the proposed trail and also its main characteristics to guide construction.

Based on the data obtained, an environmental interpretation of the trail was also drawn up. Because it is in a Permanent Protection Area (PPA), we chose the theme of water resources. What's more, this topic is very much in evidence in today's media, mainly because of the negligent way in which this natural resource is being treated in the current social context, even though it is extremely important for the maintenance of life in the most diverse ecosystems. It is proposed that the trail be self-guided, using the methodology of interpretive panels to communicate to visitors the relevant information on each section of the trail.

An important feature of the trail is its proposal to be inclusive, involving people with reduced mobility, such as wheelchair users and the visually impaired. To this end, it is planned that its floor will be suspended from the ground and that it will contain interpretive signs in Braille.

2. OBJECTIVES

2.1. General:

Propose the creation of an interpretive trail based on the environmental attributes and profile of FEENA's visitors, considering it as an environmental education tool, with the aim of contributing to the Unit's Management Plan, taking into account the need to conserve the area.

2.2. Specifics:

- Identify the most suitable area in the study area for the construction of an interpretive trail, taking into account environmental and socio-cultural aspects;
- Know the profile of the visiting public, to help design the interpretive trail to meet the demand and/or expectations of visitors;
- Develop environmental interpretation activities for the trail, as an instrument to promote environmental education.

3. THEORETICAL BASIS

3.1. Conservation Units

Within the current context of environmental degradation, caused mainly by the current unsustainable pattern of production and consumption (AGENDA 21, 1992), Conservation Units are internationally considered to be one of the most successful measures for conserving biodiversity, according to the Union for the Conservation of Nature (IUCN). Created in 1948, this institution, now known as the World Conservation Union, plays a fundamental role in guiding the creation of conservation units around the world (MILANO, 1993 apud VASCONCELLOS, 1998, p.4).

The first forms of protection of natural areas took place in the Middle Ages, with the aim of safeguarding fauna for hunting by the nobility. The initial objectives were to preserve fauna, flora and scenic beauty. Over time, other objectives were incorporated. After the Industrial Revolution, due to the exhausting routine of workers, parks or protected areas also became places for public use and recreation (verbal information)[3] . Nowadays, protected areas serve as a refuge from urban life by seeking closer contact with nature, which has been lost with urbanization.

The practice of creating legally protected natural areas has been consolidated worldwide since the creation of the pioneering Yellowstone Natural Park in the United States in 1872. "The idea of a 'park' soon spread to other countries such as Canada in 1885, New Zealand in 1894 [...]; Argentina in 1903, Chile in 1926, Ecuador in 1934, Brazil and Venezuela in 1937 (IUCN, 1975 apud VASCONCELLOS, 1998, p.4). Following this practice, Brazil created its first park, Parque do Itatiaia, in 1937. Although the proposal to create this park dates back to the end of the 19th century, it was only instituted after the implementation of the Forest Code in 1934, which provided a legal basis for the creation of UCs.

Debates on biodiversity conservation and sustainable development gained momentum at the 1972 Stockholm Conference (1ª World Conference on Man and the Environment) and were strengthened at the United Nations Conference on Environment and Development (UNCED), held in Rio de Janeiro in June 1992 (MMA).

This meeting established the Convention on Biological Diversity (CBD, 1992), whose pillars are, according to Article 1 (Objectives) of Decree 02/92, "the conservation of biological diversity, the sustainable use of its components and the fair and equitable sharing of the benefits arising from the utilization of genetic resources" (BRASIL, 1992). According to Leuzinger ([20--?], p.3) in this important legal instrument "the creation of areas

---

3 Information provided during the Introductory Course to the Management of Conservation Units, in Sao Paulo, in December 2013.

was established as one of the best forms of *in situ* conservation. As a signatory, Brazil has produced a series of documents aimed at fulfilling its commitments."

In terms of legal aspects, Brazil passed Law 9.985/00 in 2000, known as the National System of Nature Conservation Units (SNUC), which systematizes the rules and establishes criteria for the creation, implementation and management of Conservation Units. In 2002, Decree 4.340 was created to regulate the SNUC. The law defines a Conservation Unit as a territorial space specially protected by the public authorities:

> [...] territorial space and its environmental resources, including jurisdictional waters, with relevant natural characteristics, legally instituted by the Public Power, with conservation objectives and defined limits, under a special administration regime, to which adequate protection guarantees apply (BRASIL, 2000).

According to the Caderno de Educaçao Ambiental (SÂO PAULO, 2009b), PAs contribute to the protection of biodiversity by perpetuating the diversity of species and genes in order to maintain the ecological stability of the area. They protect the ecological systems and processes that are vital to the quality of life and the balance of the environment, by improving water quality, reducing erosion processes and the consequent silting up of water bodies, controlling rainfall patterns and enabling the recovery of degraded areas. It also preserves the region's history and culture. It also provides a place for scientific research and environmental education.

Article 7 of the SNUC divides Conservation Units into two groups, based on their specific characteristics: Integral Protection Units and Sustainable Use Units. FEENA falls into the group of Sustainable Use Units, i.e. direct exploitation of part of the natural resources is allowed in a sustainable way, combined with nature conservation. This group is subdivided into seven categories: Environmental Protection Area, Area of Relevant Ecological Interest, National Forest, Extractive Reserve, Fauna Reserve, Sustainable Development Reserve and Private Natural Heritage Reserve. FEENA is part of the National Forest group, defined as:

> Art. 17: The National Forest is an area with forest cover of predominantly native species and its basic objective is the sustainable multiple use of forest resources and scientific research, with emphasis on methods for the sustainable exploitation of native forests (BRASIL, 2000).

Since it is a PA managed by the state, it is called a State Forest, since according to the law, article 17, "§ 6° The unit of this category, when created by the State or Municipality, will be called, respectively, State Forest and Municipal Forest." (BRASIL, 2000).

Although public visitation of the Forest category is not a basic objective in the definition of the Law, the third paragraph refers to activities aimed at public use: "§ 3° Public visitation is permitted, subject to the rules established for the management of the unit by the body responsible for its administration."

(BRASIL, 2000, p.).

3.2. Management Plan

The SNUC defines that each Conservation Unit must have a Management Plan, an environmental planning and land management instrument that is essential for the unit's management to achieve the objectives set for a Conservation Unit. This document establishes the zoning of the area, which spatially organizes the areas of the UCs, with each zone having specific rules and different degrees of protection and use, taking into account environmental and cultural specificities with a view to conservation.

FEENA has been divided into seven management zones: Historical-Cultural Zone;

Recovery Zone; Forest Management Zone; Conflicting Use Zone; Public Use Zone; Special Use Zone; Conservation Zone, each located on map 1.

The trail, which has already been proposed in a Course Conclusion Work (TCC) by Andréa Soares Pires (2001), is located in the Forest's Public Use Zone.

Map 1- FEENA Zoning

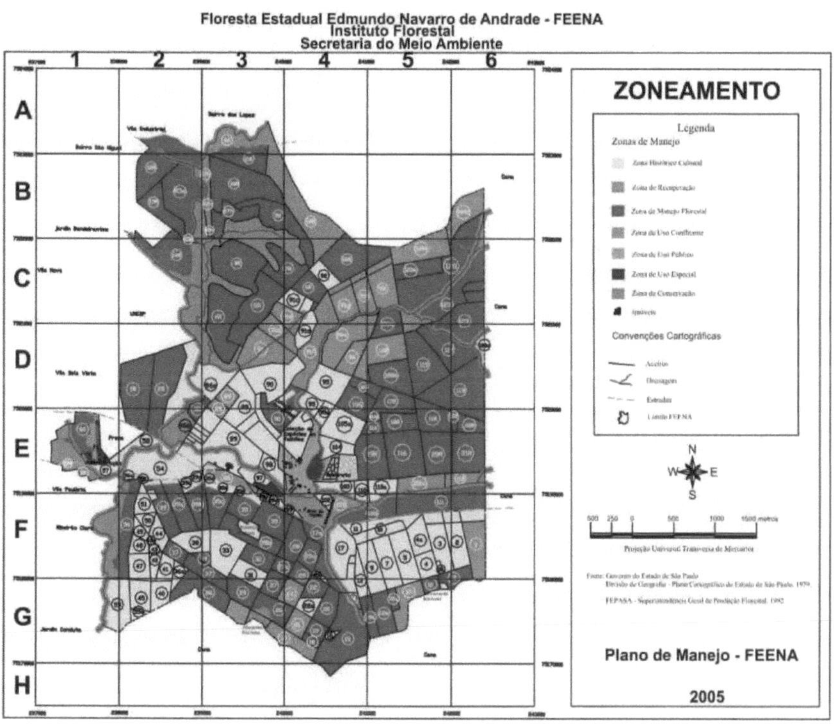

Source: Sao Paulo (2005).

According to the unit's Management Plan (2005), the public use area should be kept as close as possible to the natural environment, but with infrastructure that meets the needs of visitors. It currently includes historical and cultural heritage sites such as: Solar Navarro de Andrade, Eucalyptus Museum, Santo Antônio dos Eucaliptos Church, Casarao Amarelo, along with visitor service areas such as the Visitor Center, Auditorium, Social Center, as well as the 9 km, Health and Collection trails.

The Management Plan is also responsible for presenting the guidelines for the programs that make a PA work. These include Inspection Programs, Public Use Programs, Recovery of Degraded Areas, among others.

The implementation of new trails is foreseen as an expected result of the Nature Interpretation Program, which has the following general objectives:

> To provide visitors with closer contact with the ecological, cultural, historical, archaeological and educational characteristics of FEENA, leading them to an understanding of their interaction with the environment and sensitizing them to participate in the conservation of the Unit (MANAGEMENT PLAN, 2005, p.188).

Inevitably linked to the Nature Interpretation Program is the Environmental Education Program, which aims to:

> To provide visitors and the community with knowledge, values, skills and experiences that provoke changes in attitudes; making them able to act individually and collectively to solve local, regional and global environmental problems (MANAGEMENT PLAN, 2005, p.181).

Another program that should be considered in conjunction with the two above is Public Use.

> Among the various management programs, public use programs aim to integrate the community with the natural area, raise critical awareness of the need to conserve natural, cultural and historical resources and the value of Conservation Units, as well as encouraging their participation in the management and protection of these areas (MENDES, SOUZA, TABANEZ, 2007, p. 173).

Although there is a division into several programs, when designing activities in PAs, there is no way to separate them, since the objectives of the three programs mentioned are necessary. Vasconcellos (1998) contributes by bringing up the thoughts of Wood and Pàdua (1990) on the value of programs when well implemented for the development of educational actions:

> These programs should enable a deeper understanding and appreciation of the natural and cultural resources of the areas visited, and should also serve as a vehicle for changes in behavior, with a reorientation of habits, attitudes and values (WOOD and WOOD, 1990; PADUA, 1991 apud VASCONCELLOS, 1998, p.9).

In establishing priority lines of research, the unit's Management Plan (2005, p.329) highlights the following as a subsidy to the Public Use Program: "Development of interpretive and/or computerized methods for the Visitor Center, Eucalyptus Museum and trails;". In this way, this work contributes to the Management Plan by proposing an interpretive activity for the trail.

3.3. Environmental Education

The aim of a conservation area goes beyond conservationism. It also seeks to consolidate scientific research practices that are in tune with the socio-environmental reality of the place where the park is located. Together with the need to develop environmental education activities, in accordance with the socio-environmental characteristics of the area and the profile of its visitors, which enable the creation of links between the UC, visitors and the surrounding community, thus stimulating participation and appreciation in the maintenance and protection of the Forest.

With regard to the history of Environmental Education (EE) and its inclusion in the formal field of education, Patricio (1991) and Medina (1997) point out that:

> Environmental education emerged as a response to the environmental movement of the 1970s, for which the violation of ecological principles had already reached such a point that, at best, it threatened the quality of life and, at worst, in the long term, put the possibility of humanity's own survival at stake. Since then, environmental education has been considered as a field of pedagogical action, acquiring international relevance and validity (PATRICIO, 1991; MEDINA, 1997 apud VASCONCELLOS, 1998, p.11).

In order to show the main milestones in the development of EE in the world and in Brazil, the following is a brief history of key events:

S 1972 - United Nations Conference on the Human Environment (Stockholm Conference, Sweden), the importance of educational action in environmental issues was defined for the first time;

J 1975 - 1st International Seminar on Environmental Education (Belgrade Meeting, Yugoslavia), the first world meeting dedicated specifically to the subject. It gave rise to the "Belgrade Charter", which sets out the goals and principles of environmental education. It consolidated the International Program for Environmental Education - PIEA.

J 1977- 1st Intergovernmental Conference on Environmental Education (Tbilisi, Georgia), the final document called the "Tbilisi Declaration" was a starting point for the implementation of National Environmental Education Programs, and defined the objectives of EE.

J 1988 - Brazilian Constitution contemplates Environmental Education.

J 1992 - United Nations Conference on Environment and Development (UNCED) (Rio 92, Rio de Janeiro, Brazil). Several documents were drawn up as a result: Climate Convention or Climate Change Convention; Biodiversity Convention; Declaration of Forest Principles; Agenda 21; Earth Charter. This conference also formalized the Brazilian Charter for Environmental Education. At the same time as the conference, the Global Forum met to draw up the "Treaty on Environmental Education for Sustainable Societies and Global Responsibility".

J 1994: - The National EE Program (PRONEA) is approved in Brazil, which provides for actions in the areas of formal and non-formal EE.

J 1997 - 1st National Conference on Environmental Education (Brasilia, Brazil);

J 1999 - Law No.° 9.597/99 establishing the National Environmental Policy (PNEA) is approved.

J 2009- Launch of the ENCEA booklet (National Strategy for Communication and Environmental Education in the SNUC).

In this paper, environmental education is defined according to the National Environmental Education Policy (PNEA), established by Law No. 9.795/99, as:

> [...] a process through which the individual and the community build social values, knowledge, skills, attitudes and competencies aimed at conserving the environment, which is a good for the common use of the people and essential to a healthy quality of life (BRASIL, 1999).

According to the above-mentioned law, environmental education can be carried out in a formal environment, i.e. in school education developed within the curricula of public and private educational institutions, but also in a non-formal environment, which characterizes the type of education aimed at raising awareness among the community about environmental issues and their organization and participation in defending the quality of the environment outside the school environment, such as activities that raise awareness about the importance of PAs.

Vasconcellos (1998) collaborates by defending environmental education as an instrument for reconnecting people with the natural environment:

> This distancing, which has arisen as a result of the civilization process and urbanization itself, places most people in a daily life so disconnected from the environmental reality that they can no longer perceive the consequences or effects of their attitudes on the environment [...] One of the objectives of environmental education is to re-establish people's 'connection' with their environment, whether natural or built. (VASCONCELLOS, 1998, p. 11)

With the aim of awakening in people a feeling of closeness and respect for nature, appreciating its importance in sustaining human life and the life of the planet itself, environmental education activities have become essential in the context of PAs. By understanding why nature should be conserved, people will be encouraged to defend areas set aside for nature conservation.

3.4. Environmental interpretation

When visiting a natural area, visitors generally see the place as a source of leisure and recreation;

they don't go in search of learning and end up bringing urban habits to the place (MENGHINI, 2005). People are disconnected from the natural environment and even from built environments, to the point of not realizing the effects of their own attitudes on the environment. Environmental interpretation has emerged in this context since the end of the last century, as a way of awakening in people a new way of looking at the world, creating new awareness, understanding and commitment to their surroundings, all in a pleasurable way (VAscoNcELLos, 1998).

IBAMA (2002) cites Tilden (1977), who is considered the first to conceptualize the term environmental interpretation:

> [...] an educational activity, which aims to reveal meanings and interrelationships through the use of original objects, direct contact with the resource and illustrative means, rather than simply communicating literal information (TILDEN, 1977, apud IBAMA, 2002, p.11).

Based on the principles of Tilden (1977, apud IBAMA, 2002), the

Interpretation should be:

It's enjoyable and engaging, with the means of communication differing from the formality of the classroom;

It ***is*** meaningful and relevant, being able to relate the content to something the person already knows or has experienced;

It's organized and coherent, with interrelated ideas and a beginning, middle and end;

It's provocative, instigating the visitor to reflect;

It's different, as the visiting public is very diverse;

It has a theme and a message to convey.

In this way, it can be seen that transmitting information alone is not the focus of environmental interpretation. The information is used in a captivating way in order to pique the visitor's curiosity and provoke them to rethink their relationship with their surroundings, thus revealing new meanings and relationships.

In addition, it is worth highlighting other relevant authors in defining the concept of Environmental Interpretation.

For Ham (1992), interpretation is a means of communication with an accessible language capable of bringing people closer to the natural environment and making them interact and reflect on it. Technical language needs to be translated so that it is easy for everyone to understand, following four basic characteristics: mild; pertinent; organized and thematic.

Andrade (1993 apud CERATI, 2000, p. 33) emphasizes that the aim of interpretation is to provide visitors with a greater enjoyment of their visit, leading them to a greater understanding of the environment and its interrelationships.

For Pagani [et al.] (1996, p. 32) interpretation:

> [...] It is a didactic tool, flexible and adaptable to the most diverse situations, which seeks to clarify the phenomena of nature for a certain target audience, in appropriate and accessible language, using the most varied means, promoting in this audience an understanding and respect for nature.

According to the document "Guidelines for Visits to Conservation Units", drawn up by the Ministry of the Environment in 2006, one of the guidelines for environmental interpretation is:

> Use the various techniques of environmental interpretation as a way of encouraging visitors to develop awareness, appreciation and understanding of natural and cultural aspects, transforming the visit into an enriching and enjoyable experience (SÂO PAULO, 2006, p.17).

3.5. Interpretive trails

Due to the need to travel to other places, trails were built by human beings for various purposes, such as meeting basic needs, searching for food and water, commercial trips or religious purposes (VASCONCELLOS, 1998, ANDRADE; ROCHA , 2008). In the animal kingdom, there is the example of trails.

built by the ants as they carry leaves into the anthill, serving to cultivate the fungi that will feed them.

Over time, various changes in society have led to a change in the purpose of trails. Today they are often used as a refuge from city life, as a way of re-establishing contact with nature, which has been lost due to the very process of civilization and the unplanned urbanization seen in Brazil's big cities. In this way, they serve to contemplate the environment, appreciate physical and biological aspects and interact with nature (VASCONCELLOS, 1998, IBAMA, 2002, ANDRADE; ROCHA, 2008). According to CARVALHO and BOÇON (2004, p.24):

> The practice of walking in natural environments enables a better understanding of the environment and its interrelationships, while also sharpening the dynamics of observation, reflection and awareness of environmental issues. In addition, according to Pagani et al. (1995, apud ANDRADE; ROCHA, 2008, p.1) trails have the "objective of bringing visitors closer to the natural environment, or leading them to a specific attraction, enabling them to be entertained or educated through contact with nature".

Environmental interpretation activities will be implemented along the trail proposed in this study, providing not only contact with nature, but also explanations about the environment, forest, fauna and local natural phenomena, thus constituting interpretive and educational tools.

According to GUILLAUMON (1977), an interpretive trail is a route through a natural site that provides a closer relationship between the individual and the environment. It is an important educational tool that provides explanations about the environment (flora, fauna, geology, local history, ecological relationships, environmental protection). For Ham (1992 apud VASCONCELLOS, 1998, p.15) "a trail is

considered interpretive when its resources are translated for the visitor through specialized guides, interpretive leaflets, panels or even recordings", always as a means of developing a new field of perception.

Inevitably linked to the context of environmental education, environmental interpretation on trails has the purpose of sensitizing and making the community aware of the importance of the existence of the UC, seeking involvement and participation in actions for its conservation and enhancement. According to Mendes; Souza; Tabanez (1997, p. 20), interpretive trails: [...] offer opportunities for direct contact with the natural environment, aimed at learning and raising awareness. They also provide opportunities to reflect on values, which are indispensable for behavioral changes that are in balance with the conservation of natural resources.

From his experience with interpretive trails and nature experiences, Guimaraes (2007) notes the effectiveness of interpretive trails as follows:

In particular, in our interdisciplinary studies and practices, developed over the last twenty years on interpretive trails and experiences in nature, we can say that these experiential learning experiences are relationships of true immersion, in which we observe intense sensory and affective responses in relation to the landscape and its constituent elements, as a learning environment for an ecological view of the world (GUIMARÂES, 2007, p. 110).

Interpretive trails thus become important tools for environmental education, since their potential attractions bring human beings closer to nature, as well as making it possible to learn, raise awareness and reflect on the environment through interpretive activities.

There are two types of interpretive trails commonly classified in academic literature: guided and self-guided. Guided trails are those where a guide/monitor accompanies you on the theme of the trail in question. Self-guided trails, on the other hand, are done without a guide/monitor; the visitor walks the trail on their own. In order to approach the interpretive theme, another type of methodology is needed, with selected strategic stopping points containing panels and/or signs with the message to be conveyed or explanatory leaflets designating the stops and the explanations for each one. One of the positive aspects of the self-guided trail is that the information is always available to the public.

Table 1, contained in the Manual of Introduction to Environmental Interpretation, drawn up by IBAMA for the Doces Matas Project (2002), shows the negative and positive aspects of both guided and self-guided trails.

Whether guided or self-guided, the use of interpretive trails is a valuable educational strategy, as it puts visitors in direct contact with nature, encourages *on-site* learning and raises awareness of socio-environmental issues. The interpretive trail can be considered a "[...] communication bridge that connects visitors to resources and takes people to a new and fascinating world, providing new understanding, new ideas, new enthusiasm and new interests" (MENGHINI, 2005, p. 39).

Chart 1 - Positive and negative aspects of the trail modalities

QUADRO 13

Modalidade	Aspectos positivos	Aspectos negativos
Guiada	• Possibilita um contato pessoal com o Intérprete. • Possibilita responder, pessoalmente, quaisquer dúvidas. • Permite um controle mais eficaz da integridade do patrimônio. • Pelo fato do público ser guiado, o risco de degradar o ambiente é menor. • Adapta-se a condições diversas como eventos imprevistos, tamanho do grupo, etc.. • Os conteúdos e trajetos da atividade podem ser mudados, de uma temporada para outra, sem custos extras. • O nível de Interpretação é adaptável ao grupo. • A qualidade da mensagem é considerada alta, em função dos dois últimos itens.	• A qualidade dependerá da habilidade e do conhecimento do guia. • O visitante é obrigado a acompanhar o ritmo do guia. • Com muitos visitantes, o resultado da Interpretação pode ser comprometido. • Por trabalhar com poucos visitantes, a atividade pode não ser recomendada, para locais onde a demanda é grande.
Autoguiada	• Permite ao visitante seguir em seu próprio ritmo. • Serve de orientação para pessoas que estão perdidas. • Permite o acesso, independente da existência de um guia. • Pode ser uma alternativa, para quem não gosta de participar em grupos organizados. • Permite receber um número maior de visitantes. • Indicado para aqueles pais que querem explicar, a seus filhos, aspectos de seu interesse e de seu nível de compreensão. • Pode direcionar o público, para uma área que aceita uso mais intensivo, poupando outras áreas mais frágeis.	• Não responde dúvidas. • É suscetível a vandalismo. • O custo de manutenção pode ser alto. • O local pode ficar congestionado, se utilizado para outros fins não interpretativos. • Não é suscetível a adaptações. Restringe-se ao estabelecido nos roteiros. • Deve ser dirigido a um visitante "médio", já que não pode satisfazer às demandas dos diferentes níveis de conhecimento, dentro de um mesmo grupo.

Source: IBAMA (2002).

4. MATERIAL AND METHODOLOGIES

4.1. Bibliographic and cartographic survey

Firstly, a bibliographical survey was carried out on the subject. To this end, library databases were searched for the following topics: environmental education, environmental interpretation, interpretive trails, conservation units, FEENA. We read books, theses, dissertations, term papers and magazines on the topics mentioned above.

One of the TCCs researched proposed a system of trails for FEENA, with indications of new trails, one of which was the Jambolao Trail, the focus of this work. However, the author presents the trail proposal in summary form, with its description containing only the name of the trail, approximate location, as well as some physical characteristics, without detailing and deepening the studies on the trails themselves, not presenting the necessary data if the implementation of any trail were to be carried out.

The *Internet* was used to search for laws relating to the topics mentioned above, as well as manuals published by the government that guide work in the area of environmental education and management of Conservation Units. We also searched the Forestry Foundation's *website* and carefully read the Unit's Management Plan to obtain data on physical characteristics such as climate, vegetation, hydrography, among others, as well as socio-economic and cultural characteristics.

In December 2013, the author of this paper took part in a course at the Horto Florestal in São Paulo, organized and run by technicians from the Forestry Institute (IF) entitled: "Introduction to the Management of Conservation Units". It lasted one week and covered topics of interest to the present work, such as: the historical evolution and consolidation of the SNUC category; environmental legislation; management plans; georeferencing information in Conservation Units; environmental education; and trail management, all of which made a consistent contribution to the development of this interpretive trail proposal.

In addition, from March 2014 to the present day, the author of this research has been working as an intern in the Forest Management Department of the Planning, Development and Environment Department (SEPLADEMA) of Rio Claro City Hall. This department is located within FEENA, the result of an agreement between the City Council and the Forestry Foundation for the management of the Unit. This internship provided an opportunity to learn about the history and importance of FEENA, its physical and biological attributes, as well as the perception of the place acquired through daily interaction.

Regarding the environmental education activities already carried out at FEENA, BK employees provided documents on the route followed by the monitors, as well as explanations of the projects carried out, through an informal conversation.

4.2. Questionnaire

The questions were designed taking into account the aim of the research, which is to create an interpretive trail that takes into account not only the environmental attributes of the site, but also the profile of the visitors.

When drawing up the questionnaires, the conditions stipulated by Marangoni (2005, p. 168) were followed:

> Questionnaires for scientific research must comply with requirements such as being structured on the basis of clear hypotheses and the wording must be objective; clear; logically organized and grouped when appropriate and necessary; good graphic presentation; concise.

A questionnaire with 14 closed questions was drawn up with the aim of characterizing the form of visitation and the profile of the public visiting the Forest, as well as the visitor's relationship with the area visited (Appendix A). The model used was the questionnaire used by Vasconcellos (1998) in his research

Basic socio-economic factors of the visitors were identified, such as: income, age, schooling, among others, but also characteristics of the visit such as: frequency of visit, stay in the CU, means of transportation. As for the visitor's relationship with the site, they were specifically asked about their preferences for receiving information about FEENA, since the trail will be interpretive.

During the questionnaire, it became apparent that visitors and municipalities identify the site as Horto rather than FEENA. Therefore, where reference was made to FEENA, the official name was replaced by Horto, adapting the language to be understood by the public.

Weekends were used to administer the questionnaires in the Forest's public use area, as this is the time and place with the highest number of visitors. This stage of the survey took place over the course of a month, between the end of September and the end of October 2014. A total of 82 questionnaires were administered over four Sundays during the afternoon.

Due to the limited amount of time to carry out the survey, as it was a course completion project, it was not possible to administer a larger number of questionnaires. According to official FEENA data, around 15,000 people visit the Unit every month. Therefore, in order to effectively profile the visitors, a larger sample than the one collected would have been necessary. However, with the data acquired in this study, it is possible to arrive at a coherent approximation within the context of this scientific work.

4.3. Mapping and characterization of the trail

The trail was mapped through on-site fieldwork. At this stage, the trail was marked out point by point using a Garmin eTrex Venture HC GPS (Global Positioning System) receiver, which provides information from twelve satellites on local geographic coordinates. The GPS opted for the Route tool to mark the points. As the trail is approximately 217 meters long, it proved feasible to mark several nearby points in order to get the most accurate route possible. To edit the trail map, Garmin's MapSource software was used to generate a georeferenced location map of the FEENA trail, based on the satellite image provided by Google Earth. The map was used to indicate the places that serve as reference points for locating the trail using Corel Draw X3 software.

ArcGIS 9 (ArcMap 9.3) was used to include the location of the trail in the physical maps already available in the Management Plan.

4.4. Preparation of interpretive material for the trail

The map of the location of the interpretive signs was drawn up using the same procedures as those described in section 4.3 on mapping the trail. What changed was the point collection process, which took place only at specific points where the interpretive and informative signs were proposed to be located.

4.5. Field work

4.5.1. At FEENA

Three field studies were carried out in the area where there was already a proposal to establish the trail, which is in line with the area previously mentioned in PIRES' (2001) thesis. As the trail is not yet in place, it was necessary to enlist the help of specialized staff from Rio Claro City Hall along the way to open up the path through stretches of closed forest. A fourth field study was carried out in July on a different trail (Coleçao Trail) to the one proposed in this study, in order to observe and model the interpretive signs.

The first fieldwork was carried out in June, with the aim of getting to know the site (photo 1). In this field, general observations were made of the area with the aim of perceiving and feeling the environment, surveying superficial characteristics of physical, floristic and faunal aspects. The site was very pleasant, with a slightly hilly topography and a humid environment due to its proximity to the Santo Antônio stream.

Although the riparian forest has been altered and contains individuals of exotic species such as Eugenia Jambolana and Ecaliptus citrodora, it is a pleasant and beautiful landscape. The fauna included fish, ants, lizards, bees and bird species. It should be noted that it was not possible to identify the species. The observations made it clear that it would be possible to cross the stream three times along the trail, in order to make the trail more dynamic and make better use of the space.

Photo 1 - Jambolao trail

Source: Panagassi (2014).

The second piece of field work was carried out in August to mark points using a GPS receiver in order to trace the route of the trail, according to the specifications already detailed in the section on mapping the trail.

The third piece of fieldwork was carried out in order to mark the GPS points of the places where the interpretive panels will be installed. It took place in September.

4.5.2. In the Botanical Garden of Sao Paulo

On March 15, 2014, a field trip was made to the Botanical Garden of Sao Paulo, located in the Agua Funda neighborhood in the city of Sao Paulo, in order to see and observe the Trilha da Nascente (Spring Trail), a suspended trail on site. The trail is 360 meters long, with a linear route lasting an average of 30 minutes, with three stopping points for contemplation and observation and benches for resting, with easy access for people with reduced mobility, such as wheelchair users and the elderly (photo 2).

Photo 2 - Spring Trail

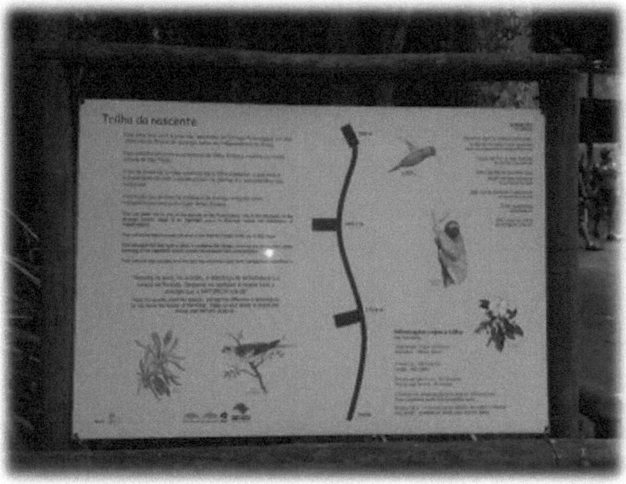

Source: Panagassi (2014).

It runs along the banks of the Pirarungàua stream, a tributary of the Ipiranga stream, and ends at the source of the same river. Located in a natural forest remnant of the Atlantic Rainforest, the trail initially goes through an area of regenerating forest, then passes through a more preserved area of considerable biodiversity.

It was built entirely suspended from the ground, using reforested wood (photo 3), in order to avoid compacting the soil, trampling plants and silting up the springs.

Photo 3 - Suspended Wooden Trail

Source: Panagassi (2014).

Also with the aim of producing the least possible impact, it was possible to observe that the removal of trees was avoided by incorporating those that would be in the path of the trail, as shown in photograph 4.

Photograph 4 - Incorporation of Trees into the Trail

Source: Panagassi (2014).

Along the way, there are interpretive signs on many parts of the trail, giving information about animals found in the area, such as howler monkeys (photo 5), explanations about conservation units and their importance, and the management of degraded areas, among other subjects.

Photo 5 - Interpretive sign

Source: Panagassi (2014)

The ease of movement provided by the suspended trail with no slope gives opportunities to small children, babies in strollers, the elderly, wheelchair users, among others, who are prevented from doing other types of more physically demanding trails. The trail has thus proved to be an integrative project, providing social inclusion by allowing people with reduced mobility to be included in the context of nature conservation spaces, while at the same time conserving nature and providing environmental education activities.

5. RESULTS AND DISCUSSION

5.1. Physical and socio-economic characterization of the study area

FEENA is located at geographic coordinates: Latitude 22° 25' S and Longitude 47° 33' W; in the eastern part of the urban area of Rio Claro, a municipality in the interior of the state of São Paulo, in the Corumbatai river basin (map 2). It currently covers an area of 2,230.53 hectares

Map 2 - FEENA location within the Corumbatai Basin, SP

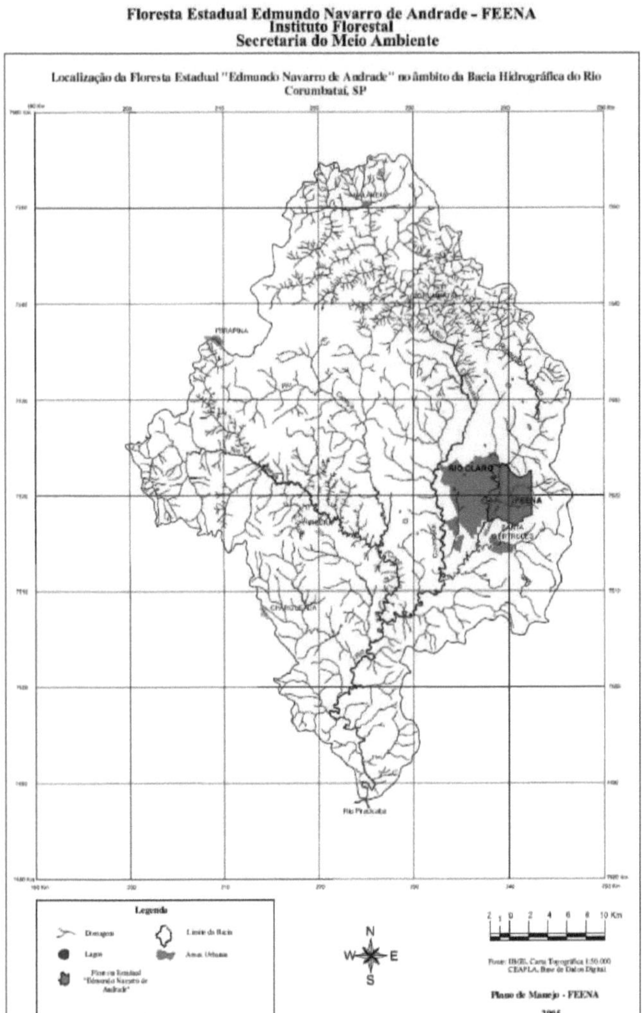

Source: Sao Paulo (2005).

▪ History

Considered the birthplace of eucalyptus in Brazil, FEENA's history is intertwined with the creation of the city of Rio Claro. At the beginning of the 19th century, Gertrudes Euphrosina de Aguiar, who came from a family with wealth linked to tropeirismo, married Antônio Paes de Barros, from a family that owned land for growing sugar cane. After their marriage, Paes de Barros acquired land in the region where FEENA is today, to plant coffee in 1825. In 1830, he founded the Santo Antônio and Sao Joao farms on part of the land where FEENA is located today. He later became the First Baron of Piracicaba and, with his political and economic influence, helped to raise the town of Morro Azul, now Rio Claro, to the status of a city (ALVES; PERINOTTO, 2012).

As coffee production declined, Paes de Barros switched to sugar cane and began renting out his land to the point of selling it. In 1909, Companhia Paulista de Estrada de Ferro acquired 580 alqueires of land in Rio Claro belonging to the Santa Gertrudes farm.

These lands were used for reforestation with the aim of producing material for the construction and maintenance of the railroad line that transported coffee production from the interior to the Port of Santos, since the demand for raw materials was high and huge areas of native forest had already been cleared, as the authors point out:

> The railroad had not been affected until 1890. However, with the excessive rise in the cost of coal, the immediate solution was to use firewood, which was still plentiful. It only took 30 years, however, for the forests on the banks of the railroad to be completely decimated (DAHER et al., 2009 p. 4).

In 1916, Companhia Paulista acquired two more farms, Santo Antônio and Cachoeirinha, totaling 720 alqueires for reforestation. With this in mind, in 1903, Mr. Conselheiro Antônio da Silva Prado, director of the Companhia Paulista de Estrada de Ferro, hired Edmundo Navarro de Andrade, a recent graduate in Agronomic Engineering from the National School of Agriculture in Coimbra, to research the best species to replace coal in locomotives and provide wood for railroad sleepers. After five years of studies and experiments, he found that species of the Eucalyptus genus, originally from Australia, were the most suitable for reforestation and the environmental conditions of the state of Sao Paulo.

After acquiring new land, Navarro and his wife moved from the Jundiai Garden to the Rio Claro Garden, which became the headquarters of Companhia Paulista's Forestry Service. In Rio Claro, he developed research activities focused on entomology, fauna, flora, genetic improvement of eucalyptus, nursery and planting techniques involving adaptation and growth studies of 144 exotic species of the Eucalyptus genus. Navarro died in 1941, leaving an immense legacy for forestry.

During the 1970s, the railroads were nationalized, and Companhia Paulista became a state-owned company, becoming Ferrovias Paulistas S.A. (FEPASA); the Rio Claro Garden was then administered by FEPASA.

In 1974, the Council for the Defense of the Historical, Artistic, Archaeological and Tourist Heritage of the State of São Paulo began the process of listing the building, and in 1977, Condephaat decided in favor of listing it.

Decree No. 2.502 of February 18, 1998 authorized the incorporation of FEPASA by Rede Ferroviària Federal S.A.. - RFFSA; with a view to its privatization, an auction was held for the concession of the railroad, however, assets such as the "Edmundo Navarro de Andrade" Forest Garden were not part of the auction, and were left under the responsibility of the São Paulo State Department of the Environment, through the Forestry Institute.

By decree 46.8 19 of June 1, 2002, the Horto Florestal was transformed into a "Sustainable Use Conservation Unit", included in the category of State Forest, and renamed the Edmundo Navarro de Andrade State Forest. It is currently managed by the Forestry Foundation.

- **Weather**

According to the Koeppen classification, the climate is Cwa, a subtropical climate with dry winters and hot summers. The average annual temperature is 20.6°C, with the hottest period being from September to April with averages above 22°C; and the least hot period, from May to August, with temperatures below 19°C.

Annual rainfall is 1,534 mm/year. The rainy season runs from October to March, when it rains 1,188 mm, equivalent to 77% of the annual total, with the wettest months being December, January and February. Between April and September there is the dry period, when it rains 346mm, the remaining 23%,

The months of June, July and August had the lowest rainfall (graph 1).

Graph 1 - FEENA Climatological Norms

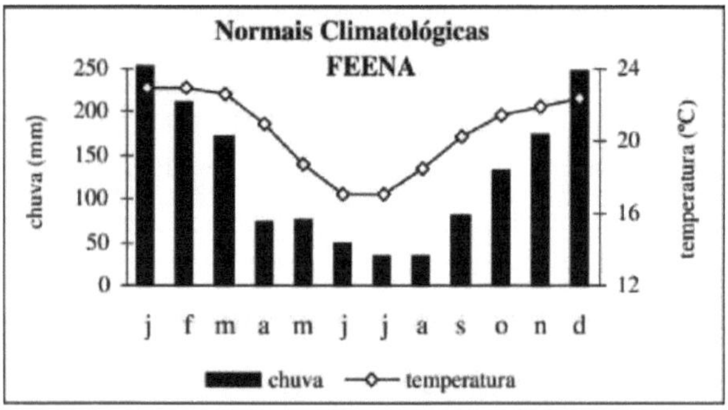

Source: Sao Paulo (2005).

According to the climatological water balance (Thornthwaite; Mather, 1955 apud SÂO PAULO, 2005, p32

[...] the annual water deficiency is only 7 mm, concentrated in July and August. While the annual water surplus is 572 mm, with the greatest concentration between October and March [...]soil conservation practices should be adopted to prevent erosion, due to water surpluses in the summer months.

Graph 2 - FEENA's Climatological Water Balance

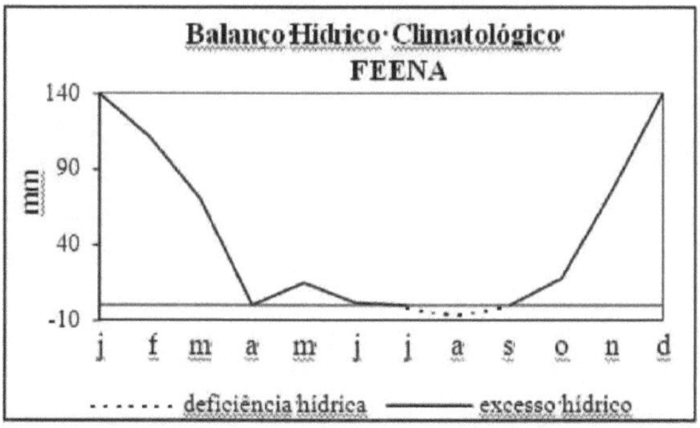

Source: Sao Paulo (2005).

■ **Geology**

The following lithologies predominate in FEENA: Rio Claro Formation, with the presence of sandstones; Corumbatai Formation, consisting of siltstones and claystones; Pirambóia Formation, with basic Mesozoic-intrusive lithologies.

Along the Ribeirao Claro canal, there is a continuous presence of Quaternary alluvial deposits made up of sands and clays. These deposits are also observed discontinuously along the Santo Antônio stream (SÂO PAULO, 2005, p.38).

The trail is located in the lithostratigraphic unit of Alluvial Deposits, formed during the Cenozoic period and composed of sands and clays (map 3).

Map 3 - Geological Characterization of the Study Area

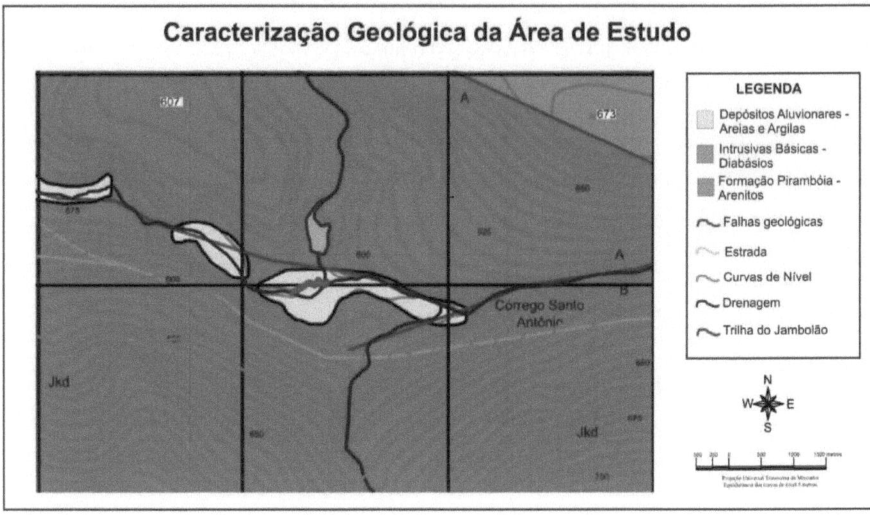

Source: Sao Paulo (2005) modified by the author (2014).

Geomorphology

FEENA is located in the São Paulo Peripheral Depression. According to the Management Plan, the area "is characterized by the presence of tabular interfluves, stepped terraces and plains, with the lowest altitudes between 550 and 600 metres and the highest between 600 and 650 metres" (SÂO PAULO, 2005, p. 33).

As far as clinography is concerned, the trail's location is dominated by slopes varying between 2% and 5%.

The following map shows that the trail is in an area of low erosion susceptibility, but it is surrounded by areas that are highly prone to erosion and is classified as an area of high erosion susceptibility (map 4).

Map 4 - Erosive susceptibility of the study area

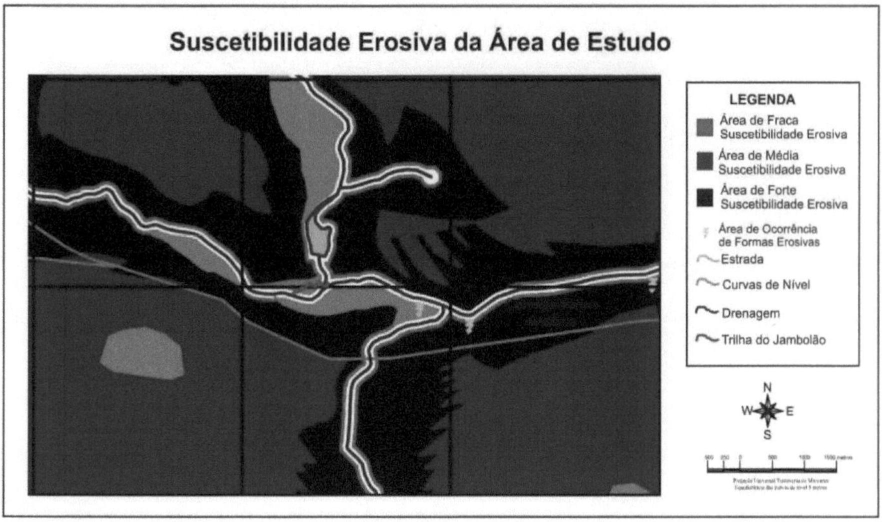

Source: Sao Paulo (2005) modified by the author (2014).

- **Pedology**

The predominant soil in FEENA is Argissolo. But there are also the following types of soil: Quartzarenic Neosols, Cambisols, Litholic Neosols and Gleissols. "All soils have a certain susceptibility to erosion, which is greater the steeper the slope. " (SÂO PAULO, 2005, p.39). As shown on map 5, the trail is located on the Gleissolo soil type, characterized by the mottled colors (greyish and/or greenish) resulting from the reduction of iron ions that occur in constantly flooded places; this type of soil commonly develops in recent unconsolidated sediments near watercourses (EMBRAPA), as is the case in this work.

Map 5 - Pedological Characterization of the Study Area

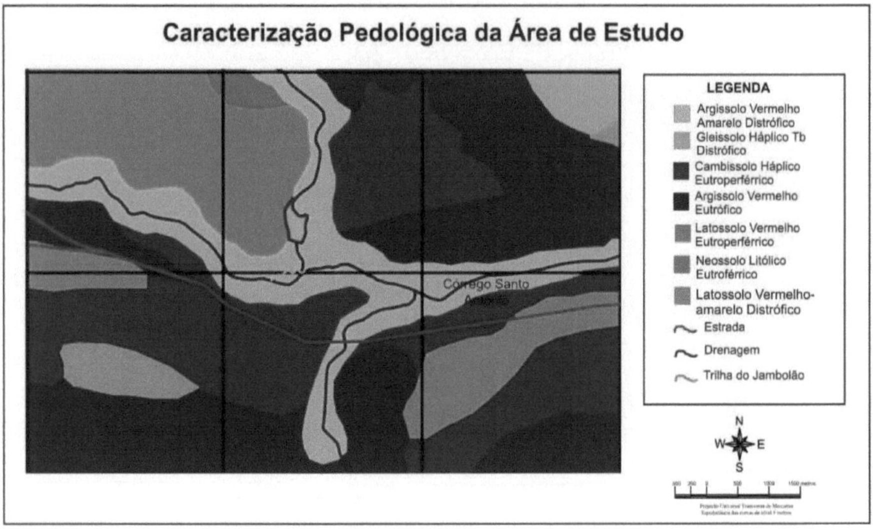

Source: Sao Paulo (2005) modified by the author (2014).

Hydrography

FEENA is located in the Corumbatai river basin, in the Ribeirao Claro sub-basin. "It is an important source of urban water supply, representing approximately 20% of the water supply for the city of Rio Claro" (SÂO PAULO, 2005 p.40). In this context, the main rivers are the Ribeirao Claro and the Santo Antônio and Ibitinga streams.

The springs of the Ibitinga stream are located in the rural area of the municipality where sugar cane is grown. According to the Management Plan:

> As already mentioned by Cunha (1998), there are high rates of soil loss in the upper reaches of the lake, above the natural dynamic, resulting from the presence of exposed soils, soil management techniques and the removal of riparian vegetation from its banks and headwaters. As a result, an intense and permanent silting process can be observed in the UC lake (SÂO PAULO, 2005, p.40).

The same is true of the Santo Antônio stream, whose headwaters are in an area where sugar cane is grown and where the riparian forest has been stripped away. "This fact can be considered responsible for the fact that this stream shows a significant reduction in its flow during periods of prolonged drought." (SÂO APULO, 2005, p.40).

There are three lakes in the FEENA area, which are the product of damming river channels: Central Lake, which originated from the damming of the Ibitinga stream; Embaùba Lake or Prona Lake, located on a tributary of the Santo Antônio stream; the water collection and treatment lake of the

Autonomous Department of Water and Sewage (DAAE), which dams the waters of the Ribeirao Claro.

According to the Management Plan, there is a process of eutrophication of FEENA's Central Lake (photo 6). According to the authors Galvao and Raduan (1982 apud SÂO PAULO, 2005, p.41):

> [...] this fact is strictly correlated to the process of damming the stream, [...] this process of eutrophication is also due, to a large extent, to the leaching of chemical products deposited in the contributing watercourses, in stretches still located outside the area of the Unit.

Photo 6 - Central Lake

Source: Panagassi (2014).

Problems such as those mentioned above, such as the removal of riparian vegetation, silting up of water bodies and the eutrophication of the lake, will be addressed on the interpretive signs along the trail.

- **Vegetation**

In the study area, the Semideciduous Seasonal Forest forms a composition of species between the Submontane Dense Ombrophylous Forest and the Cerrado. In addition to several species of eucalyptus, FEENA has living collections of gymnosperms, mainly conifers, as well as Pinus and Araucarias, and also an understorey with a varied floristic composition. Along with native species such as *Aspidosperma polyneuron* (Peroba- rosa), species of the genus *Tabebuia* (Ipê), *Jacaranda sp* (Jacarandà), *Cariniana legalis* (Jequitibà-rosa), *Caesalpiniae chinata* (Pau-brasil), *Psidium litorale* (Araça-amarelo), *Piptadenia macrocarpa* (Angico-vermelho), *Piptadenia communis* (Pau-jacaré), *Schizolobium excelsum* (Guapuruvù), *Euterpe edulismartius* (Juçara), among others.

- **Fauna**

FEENA is home to animals such as: Jaguar (*Puma concolor*), Capybara (*Hydrochaeris hydrocaeris*), Armadillo (*Dasypus sp*), Paca (*Agouti paca*), Furao (*Galictis cuja)*, Hedgehog (*Coendou sp*), Yellow-faced *caiman (Caiman latirostris*), Rattlesnake (*Crotalusdurissus terrificus*), *Boa constrictor (Boa constrictor amarali*), fish adapted to degraded environments such as Saguiri (*Cyphocharax mosdetus*) and Caborja (*Hoplosternum littorale*).Willis (2003) found more than 255 species of birds at FEENA, among them the Purple-billed Teal (*Nomonyx dominica*), the Common Skylark (*Busarellusni gricollis*), owls such as the *Asiostygius* species, the Canary (*Sicalis flaveola*), among others. Only a few species were mentioned, but there is a long list of species in the Management Plan that inhabit FEENA.

- **Socio-economic aspects of the surroundings**

Based on the data presented in the Management Plan on the socio-economic situation of the surroundings, the city of Rio Claro is also considered in this work as the surroundings of FEENA.

According to data from the 2010 Census (IBGE, 2010), Rio Claro had a population of 186,253 inhabitants. Considering that the area is 498.422 Km^2 , the demographic density was approximately 373 inhabitants per km2. There were 90,687 men (48.7% of the total) and 95,566 women (51.3% of the total).

The Municipal Human Development Index (MHDI) was 0.803. The total number of employed people is 75,091 and 49,978 people were attending nursery school or school at the time of the census. Of the total population, 89.5% are literate, i.e. 166,684 people. The average monthly income per capita in rural areas is 500.00 reais, while in urban areas the figure rises to 733.33 reais.

5.2. Characterization of Environmental Education already carried out at FEENA

The environmental education activities carried out at FEENA are provided by the company BK Consultoria Ambiental, which is outsourced by the Forestry Foundation to carry out environmental monitoring work.

One of the projects developed by the company is "Places to learn: the school goes out of the school" in partnership with the Foundation for the Development of Education - FDE, which offers free guided tours to students from the state public school system, from elementary school (after six years) to secondary school and the EJA.

For elementary school pupils, the idea is to address issues already discussed in the classroom in relation to environmental conservation, thereby making them aware of the need to change their attitudes in order to establish a more sustainable society. To this end, the students are welcomed at the Experience Center to watch a film about FEENA and then walk along part of the Health Trail with a monitor, awakening their environmental perception of the place.

The environmental education activity, aimed at elementary school students from cycle II to secondary school and the EJA, focuses on presenting the historical and cultural content of FEENA through a guided visit to the Eucalyptus Museum and then to the Health Trail, where topics related to the environment are covered in order to raise awareness among visitors.

Through the historical-cultural and environmental education project entitled "Estaçao Turismo" (Tourism Station), run by the Rio Claro City Council, through the Tourism, Education and SEPLADEMA departments, in partnership with the Ràpido Sao Paulo bus company, together with the Energy and Sanitation Foundation and BK, environmental monitoring work is also carried out in the forest for municipal schools and Social Assistance Reference Centers (CRAS). In this project, students from municipal schools and CRAS participants are taken to the city's tourist attractions, including FEENA. There, the students go on a guided tour, following the same route and objectives set for students from the state school system. In 2013, the project brought 2,480 visitors to FEENA, through 97 tours.

Another way of providing environmental monitoring activities is by scheduling groups that are open to all visitors.

The interpretive trail proposed in this work could be used as a complementary activity to these environmental education activities *already* carried out at FEENA.

5.3. aracterization of FEENA's existing trails:

There are three trails already in place at FEENA, the 9Km Trail, the Collection Trail and the Health Trail, located as specified on map 6:

Map 6 - FEENA trails

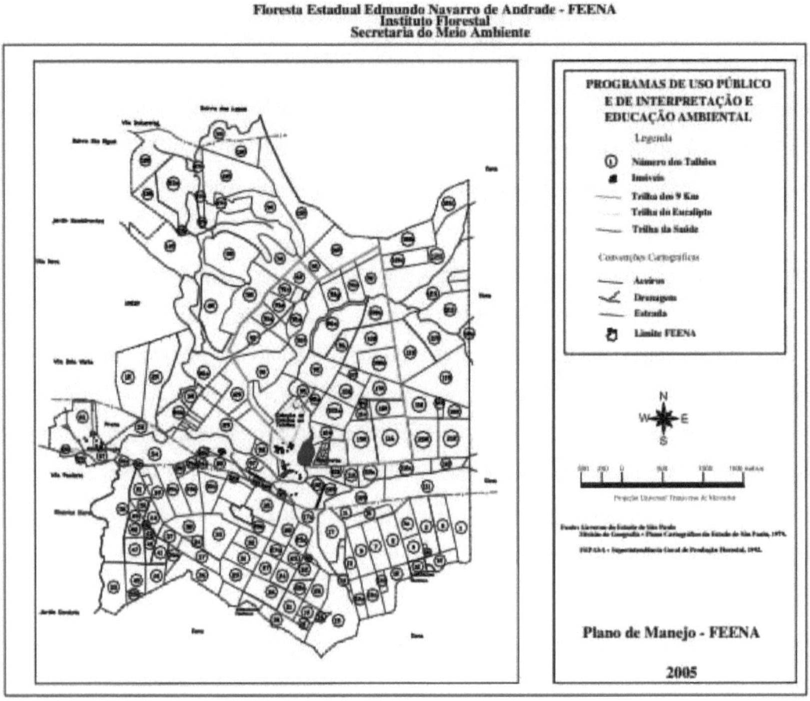

Source: Sao Paulo (2005).

- The 9 Km Trail: named after its 9 km length, it is of medium difficulty and is used for running and hiking. Along the way, it is possible to see several Eucalyptus and Pine trees of different ages, as well as crossing the Santo Antônio stream. There is no environmental interpretation on the trail, nor is there any kind of environmental education activity.

- Trilha da Saùde: The only one included in the "Trilhas de Sao Paulo" program, an initiative of the State Government's Environment Department. It has a low level of difficulty and is 1.3 km long. Located in the Arboretum, next to the Central Lake, visitors can contemplate native and exotic species. It was designed by two-time Olympic triple jump champion Adhemar Ferreira da Silva and contains equipment and explanatory signs aimed at physical fitness, so it can be used for physical exercise as well as for educational activities and scientific research. It has no environmental interpretation, although this has already been proposed in Piedade's (2013) thesis, which proposes an interpretation based on Waldorf Pedagogy.

- The Collection Trail: low difficulty, 1.5 km long, located within the Talhoes Collection (an area subdivided by Edmundo Navarro de Andrade into 68 talhoes, with different species of Eucalyptus mainly from Australia, with the aim of promoting comparative studies on the development of the species). The trail is self-guided and has twenty-two interpretive signs, all with a pyrographic outline in wood and painted white on the inside. The first sign, located right at the start of the trail, tells you the distance and degree of difficulty, along with a sketch of the route to be taken (photo 7).

Photograph 7 - Collection Trail Interpretive Plaque

Source: Panagassi (2014).

Immediately afterwards there is a plaque telling the story of the Eucalyptus collection (photo 8).

Photograph 8 - First interpretive sign on the Collection Trail

Moving on, there are plaques about the collection in line (photo 9) and in butchers, with a brief explanation of these two types of collection.

Photo 9 - Interpretive plaque for the Lines Collection

Source: Panagassi (2014).

There is also information on the riparian forest and wetlands (photo 10).

Photo 10 - Interpretive plaque about the Vârzea area

Source: Panagassi (2014).

The other plaques indicate the scientific name and some characteristic of different species of Eucalyptus found along the trail (photograph 1 1), as well as other trees, Pau-Ferro, Angico (photograph 12) and Pau jacaré.

Photograph 11 - Eucalyptus Species Interpretive Plaque

Source: Panagassi (2014).

Photograph 12 - Trail Species Interpretive Sign

Source: Panagassi (2014).

5.4. Characterization of the Visiting Public

The questionnaires were used to obtain the following data on the profile of visitors to FEENA, as well as the profile of the visit and the visitor's relationship with the area visited, more specifically as regards receiving information about FEENA.

In order to profile the visitors, they were asked about their gender, age group, marital status, schooling, origin and monthly income.

In terms of gender, the proportion of men and women was close. Of the 82 interviews, 43 were women, while 39 were men (Graph 3).

Graph 3 - Visitors' gender

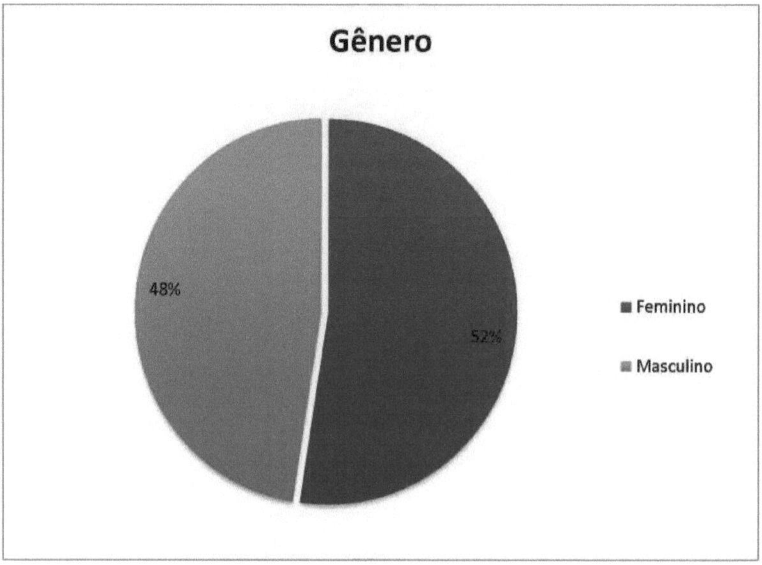

Source: prepared by the author (2014).

When it comes to the age of visitors, it can be seen from Graph 4 that the majority of visitors are aged between 20 and 40, with a significant number aged between 41 and 50. There aren't many people in the under-15 age bracket because the interview prioritized adults, but during the questionnaires, the presence of children attending the site was noted (graph 4).

Graph 4 - Visitors' age group

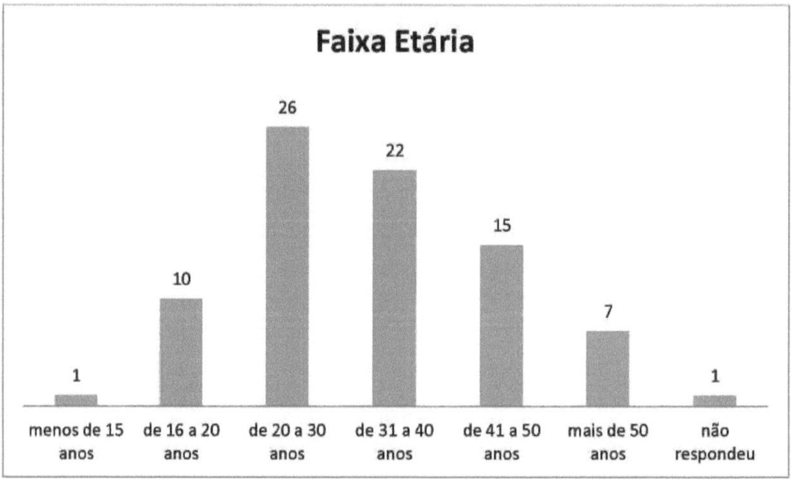

Source: prepared by the author (2014).

The majority of the sample said they were married, followed by people who were single. Of those interviewed, 38 were married, 32 were single and 11 were separated, while only one person declared themselves to be widowed (Graph 5).

Graph 5: Marital Status of Visitors

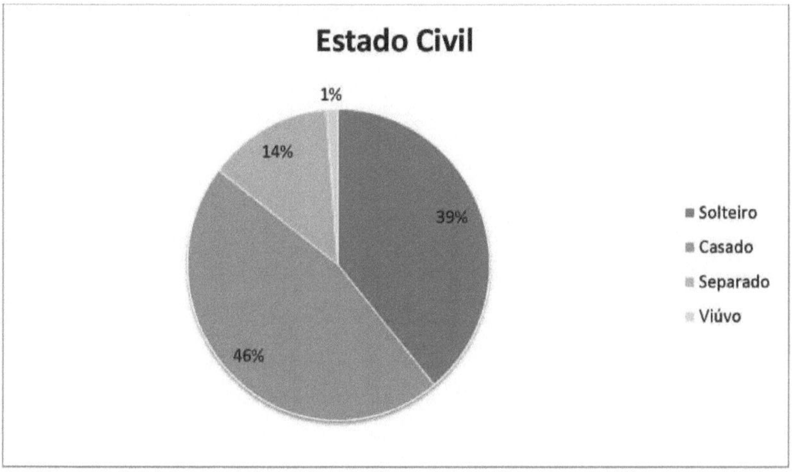

Source: prepared by the author (2014).

With regard to level of education, the most significant numbers of visitors were those with a university degree or higher, followed by those with a high school degree. There was a considerable number of individuals who had only completed primary school, showing the importance of educational panels that take into account various levels of education (graph 6).

Graph 6 - Visitor schooling

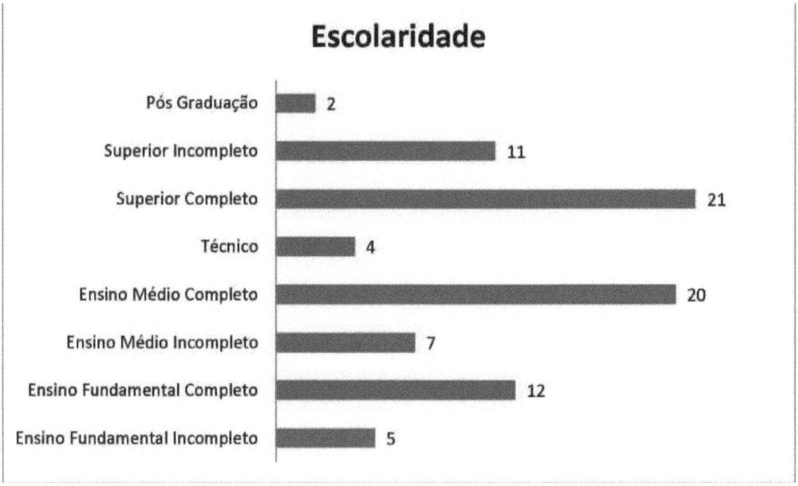

Source: prepared by the author (2014).

With regard to where the visitors came from, 52 interviewees said they were from Rio Claro itself, while 22 came from nearby towns such as Santa Gertrudes, Limeira, Cordeirópolis and Corumbatai. The rest, who totaled 8 visitors, came from other places such as Sao Paulo, Araraquara, Sao Carlos, but all live within the state of Sao Paulo (graph 7).

Graph 7 - Where visitors come from

Source: prepared by the author (2014).

In terms of monthly income, there was a predominance of individuals earning between 1 and 5 minimum wages per month, totaling 63 people, i.e. 77% of those interviewed (graph 8).

Graph 8 - Visitors' monthly income

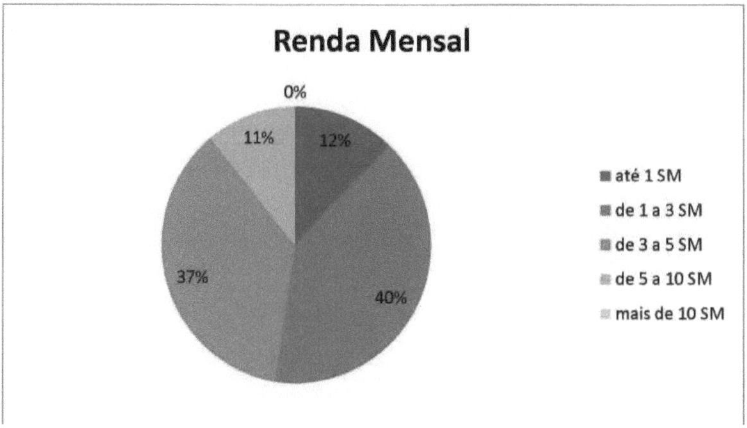

Source: prepared by the author (2014).

The following questions were asked in order to draw up a profile of visitors to the UC. To this end, visitors were asked about: the means of transportation used to get to the unit, how often they go to FEENA, who accompanies them during their visits, how long they stay in the forest and the main activity they do there.

The car was by far the most common means of transport used by visitors to get to the unit: 56 interviewees used the car, 1,3 used a bicycle, 11 used a motorcycle, while only 2 arrived at the Public Use Area on foot; no one was found to have taken the bus (Graph 9).

Graph 9 - Visitors' means of transportation

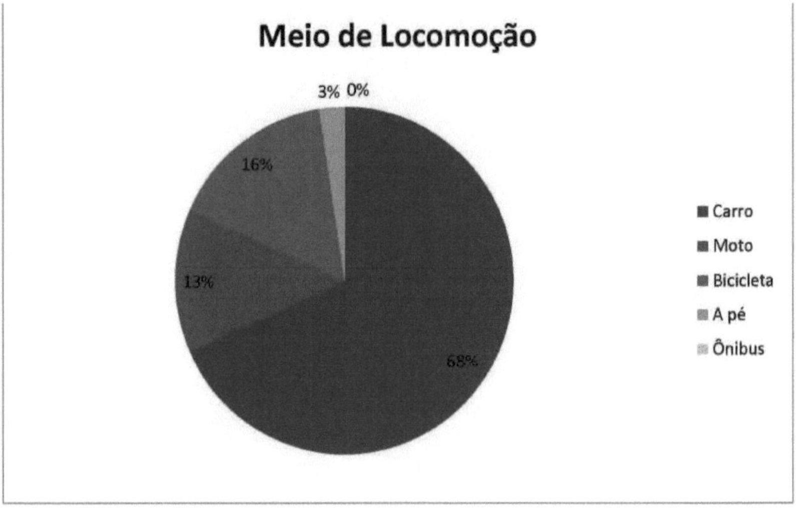

Source: prepared by the author (2014).

Regarding the frequency with which visitors come to FEENA, it can be seen that *there is* a high return rate, as 83% of those interviewed come to FEENA more than once a year (graph 10).

Graph 10 - Frequency of Visits

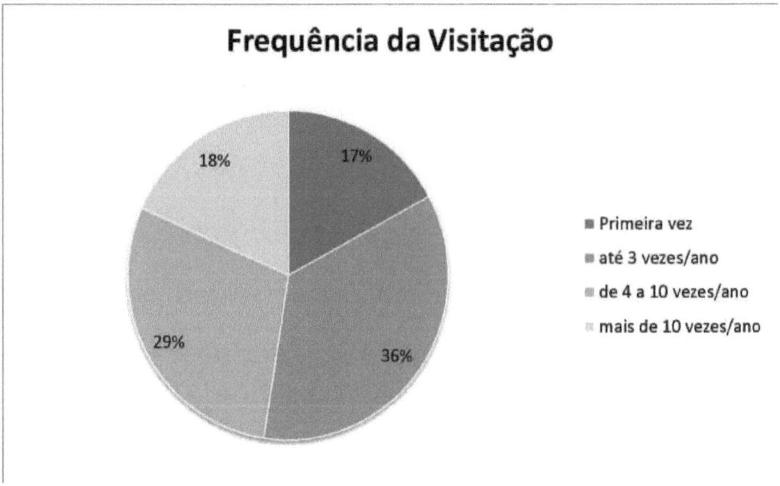

Source: prepared by the author (2014).

There is a large predominance of people who visit FEENA accompanied by their family or friends, totaling 82% of those interviewed. Although proportionally few people visit the PA alone, it is considered a considerable number of people. There were no excursions during the period in which the questionnaires were administered (Graph 11).

Graph 11 - Visit companions

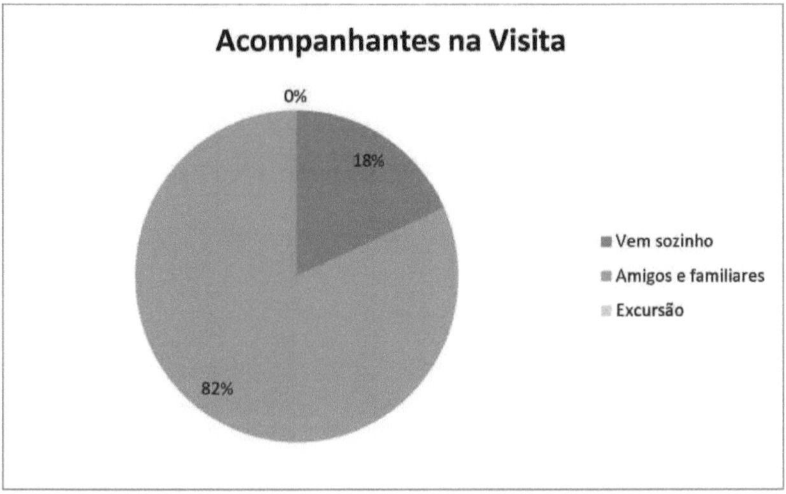

Source: prepared by the author (2014).

Most of FEENA's visitors stay at the site for up to two hours, with a total of 32 respondents. In almost the same proportion, 29 people stayed for half an hour (Graph 12).

Graph 12 - Visitor Length of Stay

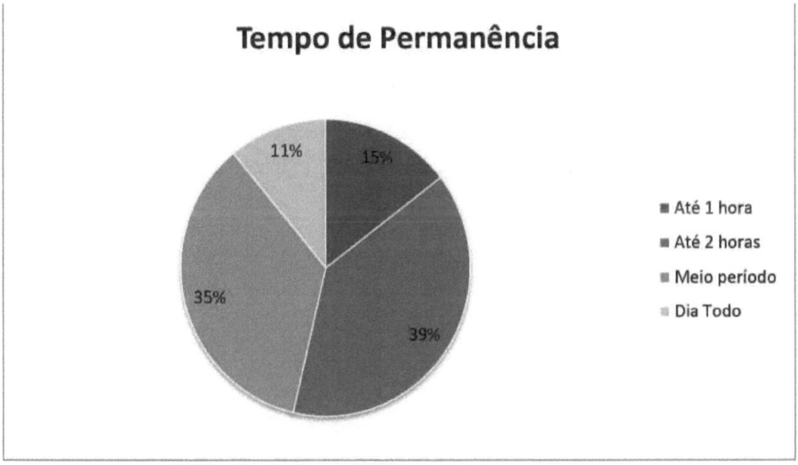

Source: prepared by the author (2014).

The main activity carried out by visitors to FEENA is walking and/or running, with a total of 39 responses to this effect. The second most popular activity is the picnic, with 22 people mentioning this activity. A visit to the Eucalyptus Museum was mentioned as the main activity by 13 people. A further 8 people mentioned other activities, the main ones being photography, yoga and cycling, despite the fact that this activity is not allowed in the unit's Public Use Area (graph 13).

Graph 13 - Main activity carried out at the CU

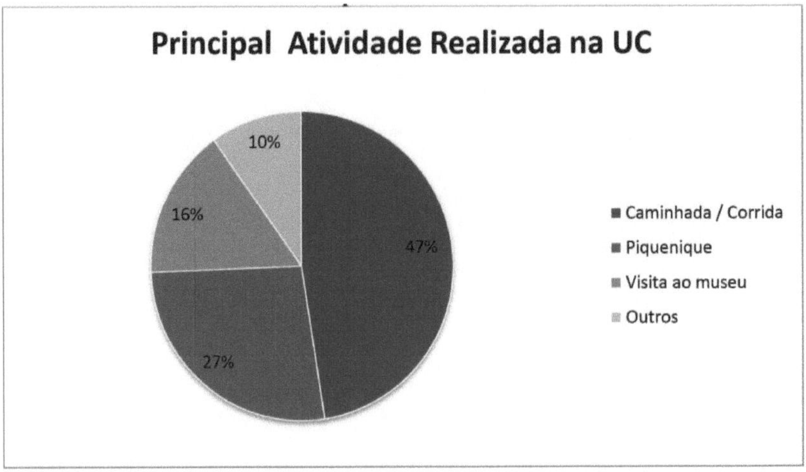

Source: prepared by the author (2014).

As this is a proposal for an interpretive trail in the forest, a survey was carried out by applying a questionnaire about the visitor's relationship with the site, specifying their preferences for receiving information about FEENA. The questionnaire asked whether the interviewee would like to receive information about FEENA; of the 82 interviewees, 73 said they would like to receive information, which is a high number of people interested in obtaining information about the site (Graph 14). Those who said yes were asked what information they would like to know about the site and how they would like to receive it.

Graph 14 - Respondents' willingness to receive information about FEENA

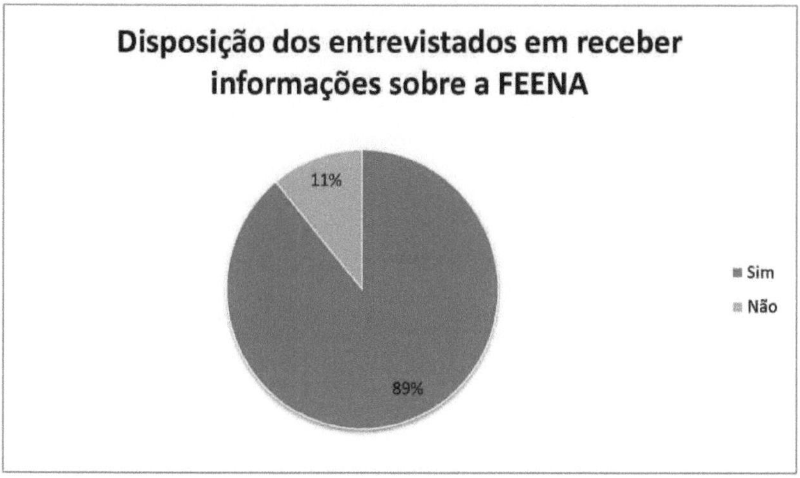

Source: prepared by the author (2014).

On the question of what visitors would like to know about the area, the vast majority of respondents indicated more than one subject. Based on this data, it can be assumed that visitors lack information about the Forest, since almost all subjects were mentioned in several questionnaires. However, it cannot be said that this is the case, as no specific question was asked about this topic, nor is the sample sufficient to draw conclusions that determine the profile of visitors to FEENA as a whole.

Information about FEENA's history was mentioned most often, appearing as a desire of 68 interviewees. Information about the local fauna was the second most cited option, appearing in 64 questionnaires. Water resources was the third most cited theme, appearing in 53 questionnaires, the fourth was flora, chosen by 49 people and lastly the importance of the area, chosen by 42 individuals. Of the total, only 5 people mentioned other subjects such as the link between the unit and the city, eucalyptus management, the conservation unit, the use of eucalyptus and general information about the garden's activities (Graph 15).

Graph 15 - Aspect to be informed about FEENA

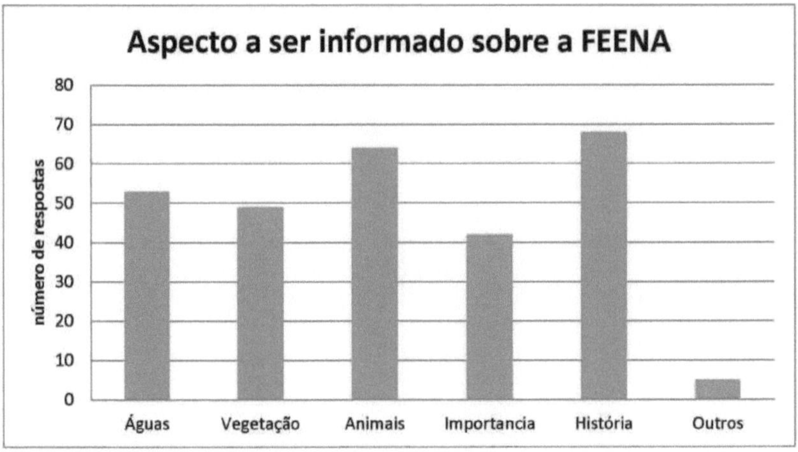

Source: prepared by the author (2014).

As with the previous question, the last point to be addressed left individuals free to choose more than one option and it was found that the vast majority ended up choosing two or more alternatives. When asked how they would like to receive information about the unit, the method most indicated by people (graph 16) was the use of signs or posters (panels), closely followed by the option of explanatory leaflets, then a guided walk and then the option for staff to provide information. The methodology of lectures appeared less frequently in the questionnaires, with almost the same number as the visitor center.

Graph 16 - How you receive information about FEENA

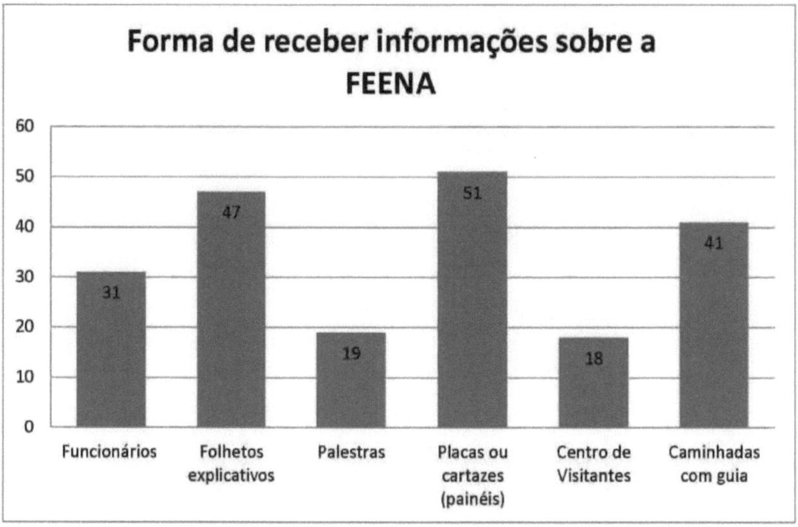

Source: prepared by the author (2014).

The questionnaires were used to draw up a preliminary profile of FEENA visitors. It is important to note that the sample collected is not sufficient to accurately define the profile of visitors, since the number of questionnaires applied was not sufficient due to lack of time.

5.5. Characterization of the Jambolao Trail

5.5.1. Location of the Jambolao Trail

According to Propst [et al] (1984 apud VASCONCELLOS, 1998), the location of interpretive trails in PAs should take into account accessibility, natural and cultural resources as well as the objectives set for the trail.

Based on this premise, it is proposed that the trail be located in FEENA's Public Use Area, next to the Horto lake, which is easily accessible to visitors, very popular, with pleasant and beautiful riverside scenery (photo 13). This site had already been proposed in PIRES' (2001) thesis for the construction of the Jambolao Trail. It was decided to leave the same name that the author had given to the trail, as well as the location, although the route that the trail will take has been modified.

Photograph 13 - Jambolao trail area

Source: Panagassi (2014).

The trail follows a stretch of the Santo Antônio stream, an area of scenic beauty. Its starting point is located at the junction of the stream in question with the waters of the Santo Antônio stream (photo 14), with the end point being the bridge near the Centro de Vivencia, over a total distance of approximately 217 meters, crossing the stream three times. The starting point is at coordinates S 22 24.873 and W 47 31.310, with an altitude of 586 meters. The end point of the trail is located at coordinates S 22 24.898 and W 47 31.402, with an altitude of 579 meters. There is little change in altitude along the route, ranging from 579 to 590 meters, so 1 1 meters.

Photograph 14 - Area where the Ibitinga and Santo Antônio streams meet

Source: prepared by the author (2014).

The exact location of the trail can be found on map 7:

Map 7 - Location of the Jambolao Trail

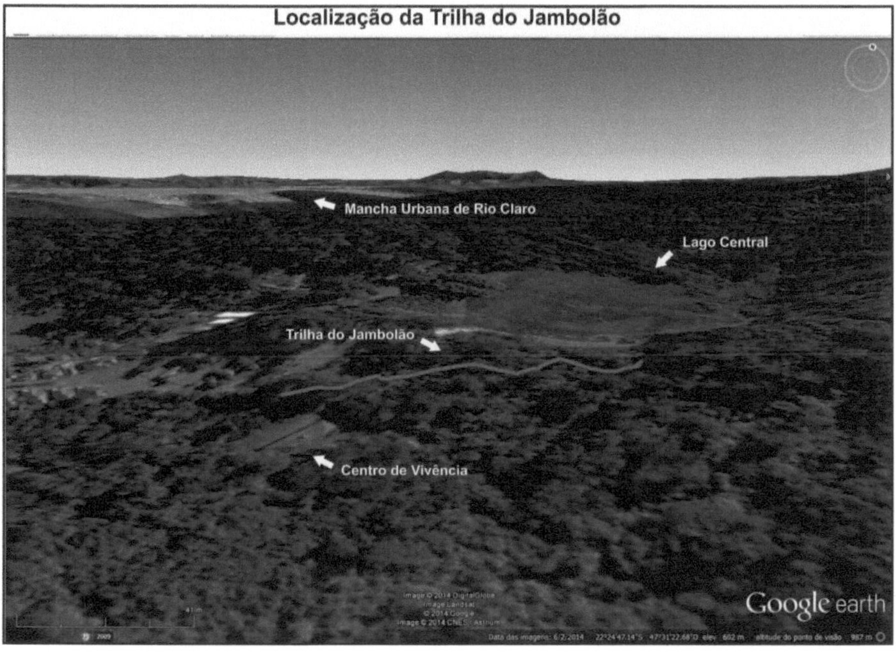

Source: prepared by the author (2014).

For the future, it is proposed to join the Jambolao Trail with the wooden walkway (photo 15) that connects the area near the bridge of the Experience Center (end of the Jambolao Trail) with the Santo Antônio dos Eucaliptos Church (photo 16), thus being able to include more interpretive points.

The footbridge was built in 1992 to facilitate access to the church for visitors to the forest, as in the summer months the rains flood the nearby meadow. The church preserves part of FEENA's history. Built by FEENA employees in 1955, it features typical European architecture.

Photograph 15 - Wooden walkway

Source: Panagassi (2014).

Photograph 16 - Santo Antônio dos Eucaliptos Church

Source: Panagassi (2014).

During the fieldwork, it was possible to see that parts of the stream bank are already being frequented by people, especially at the beginning of the trail, where frequent use is noticeable due to the lack of vegetation, making it a dirt track, as can be seen in the photograph (1 7) below.

Photo 17 - Route already taken by visitors

Source: Panagassi (2014).

5.5.2. Environmental aspects of the Jambolao Trail site

The area chosen for the trail is considered by the Forest Code (Law No. 12,651 of May 25, 2012) to be a Permanent Preservation Area (APP), that is:

protected area, covered or not by native vegetation, with the environmental function of preserving water resources, the landscape, geological stability and biodiversity, facilitating the gene flow of fauna and flora, protecting the soil and ensuring the well-being of human populations (BRASIL, 2012).

APPs have the following functions: to preserve water resources, landscapes, geological stability, biodiversity, the gene flow of fauna and flora, to protect the soil, and to ensure the well-being of human populations (BRASIL, 201 1). In addition to being within the context of a UC, the proposed trail is located in an APP, a fact that requires careful measures to protect the area.

When considering the environmental aspects, an annual water surplus of 572 mm was found, concentrated between October and March, considered to be the rainy period of the year. Due to the surplus water in these summer months, soil conservation practices should be adopted in order to prevent erosion throughout the forest. As the area chosen for the trail is a streamside area, in the wetter months the stream increases its level and floods the floodplain, making it more prone to erosion.

The riparian forest along the trail has already been altered, with several individuals of the species Eugenia Jambolana, which gives the trail its name (Jambolao Trail), due to the fact that this species inhabits several stretches of the trail. There are also various species of *Eucalyptus, Bamboo, Juçara Palm, Fan Palm, Tree Fern and Mango.* The trail needs a more detailed floristic survey of the site, as well as a fauna survey, which were not possible during the preparation of the proposal.

The soil at the site is classified as Gleissolo, typical of areas that are constantly or periodically saturated by water. The lithological unit is Alluvial Deposits, formed by sediment deposits of recent origin generally found along bodies of water. According to the Management Plan, the site chosen has a low susceptibility to erosion (map 4) and the slope varies between 2 and 5%.

Based on this environmental information, it can be inferred that although the classification indicates a low susceptibility to erosion and the slope is low, the climate, soil and geology of the site indicate a propensity to erosion, especially in rainy months. This propensity increases due to the high volume of visitors who visit the site every month (around 1,5,000/month).

Analysing the environmental diagnosis of the area and considering that the chosen sector is characterized as a floodplain of the Santo Antônio stream, it is configured as an APP. In addition, the area is seasonally flooded in the summer months as it lies on the floodplain of the stream. In view of these factors, it is proposed that the trail be made entirely suspended in relation to the ground, in order to cause the least possible impact on the environment since it also avoids compacting the soil, as well as trampling the plants. In this way, the proposal is in line with one of the specific objectives of the Public Use Program set out in the unit's Management Plan (2005, p. 143): "Its objective is to facilitate intensive recreation, leisure and environmental education in harmony with the environment".

5.5.3. Cultural and social aspects of the Jambolao Trail

According to the interviews carried out to find out the profile of the visitors, it is possible to see that the number of men and women who visit the unit is practically the same; in terms of age, there is a predominance of adults aged between 20 and 40; the majority declared themselves to be single (39%) or married (46%); in terms of schooling, there was a large number of people who are (11 people) or have already completed higher education (21 people) or who have already completed high school (20 people). The vast majority of visitors are from Rio Claro itself (63%), and have monthly incomes ranging from 1 to 3 (40%) and 3 to 5 minimum wages (37%).

With regard to visits, it should be noted that the vast majority of visitors use a car as a means of getting around (68%), and usually go to the site several times a year, accompanied mainly by friends and family (82%), staying from 2 hours to half an hour. The main activities carried out are walking and/or running (47%) and picnicking (27%).

In order to cater for this variety of people who visit the unit, taking into account that many families use the site for hiking and don't stay for long periods, a short interpretive trail would help to increase the public's leisure time, as well as attracting people to greater contact with and knowledge of the area.

As well as contributing to the lowest possible environmental impact in the area where the trail is to be built, the proposal to make it suspended provides greater accessibility to FEENA by guaranteeing an unobstructed walk for all visitors.

People with reduced mobility, such as young children, the elderly, pregnant women, the obese and the disabled, are visitors to PAs, but are often prevented from taking advantage of the leisure opportunities offered in these places. According to GONDIM (20 12, p.2) "More than half (102 million people) of our population (190.7 million in 2010) are excluded from visiting many of the PAs". There is therefore a need to plan activities aimed at this audience. With this in mind, the proposal to create a trail at FEENA will take into account those people who are normally excluded from planning the public use of protected areas.

To this end, we propose a suspended trail made from treated wood (according to the model already implemented in the Collection Trail) and leveled to soften the slope, as well as safety handrails, guide ropes and guardrails, taking into account work already carried out by Gondim (2012) in the National Forest of Brasilia and experiences gained from fieldwork carried out in the Botanical Garden of Sao Paulo.

According to the Manual for the Construction and Maintenance of Trails by the Government of the State of São Paulo (2009a, p. 23-24), "an almost flat path with gentle climbs is more appropriate in the case of interpretive trails or trails designed for people with disabilities or reduced mobility."

5.5.4. Jambolao Trail Classification

According to the classifications presented in the Manual on Trail Management drawn up by Andrade and Rocha (2008), the proposed trail falls into the following categories in terms of function, form and degree of difficulty (considering the gradation and classification of activities) described in Table 2.

Chart 2 - Classification of the Jambolao Trail

Jambolao Trail Classification	
Function	Self-Guided Trail
Form	Shortcut (start and end at different points)
Graduation	Easy
Classification of activities	Grade A: can be done without the need for physical preparation, no previous experience required
	Light trail / Short trail: distance of no more than 500 meters, requires little physical effort and specific technique, has no obstacles

Source: prepared by the author (2014).

The trail will be 1.5 meters wide in the trampling area, a width designed for the smooth passage of wheelchair users. With the inclusion of blind people in mind, a handrail with ropes will be placed to hold the person's wrist so that they can guide themselves.

5.5.5. Proposal for a self-guided interpretive trail with panels

The Public Use Program indicates the following activities for FEENA: "Adapt visual communication through signs placed in the Public Use Area, as well as trained monitors, in order to improve communication with visitors" (MANAGEMENT PLAN, 2005, p.179).

In order to contribute to the implementation of activities indicated in the Management Plan, the Jambolao Trail is proposed as a self-guided interpretive trail, as this way the information is available at all times and can reach a greater number of visitors, as well as allowing people with reduced mobility to follow at their own pace.

In order to make the trail a place of learning, it is planned to install explanatory panels (signs) along the route according to the main theme: "water resources", the message to be passed on is about the importance of maintaining good water quality in order to have a good quality of life.

Although this is a self-guided trail, there's nothing to prevent a guided tour from taking place at certain times; in this way, the interpretive panels would serve as a basis for more detailed explanations, as well as raising questions on the subject and answering queries from the guides. It would also complement the environmental education activities already carried out by FEENA with schools and CRAS.

The choice of explanatory panels as a means of receiving information about the unit was the methodology most mentioned by visitors (5 1 people), followed closely by explanatory leaflets (47 people). Explanatory panels were chosen because they don't generate waste like leaflets and because they remain at a specific point of interest, making direct reference to some aspect of the trail.

The choice of the theme of water resources is mainly in line with the context of the water crisis we are currently experiencing; it was also due to the fact that the route runs along a stretch of the Santo Antônio stream and also because the unit has problems related to water resources, such as untreated sewage dumped into the streams, the springs of the main bodies of water are located in an area of sugarcane production, and the unit's central lake is in the process of eutrophication. Another factor that contributed to the choice of topic was the interest shown by the interviewees in the subject of water, the third most frequently mentioned as one of the pieces of information that visitors would like to receive about the unit.

Each stop will have an interpretive panel presenting content linked to the main theme of the trail, consisting of a logical sequence of stopping points, with a beginning, middle and end. The content of each panel will be provided in a clear, succinct and attractive way.

Vasconcellos (1998), citing Mcintosh (1982) and Ham (1992), contributes by listing the characteristics necessary for self-guided trails to achieve their objectives and be efficient:

> In order for self-guided walks to be successful, MCINTOSH (1982) and HAM (1992) recommend that each stop, whether with leaflets or signs, as well as theme titles (not just topic titles), focus directly on a visible fact of interest, explain the facts in a quick and interesting way, relate the theme of the stop to the general theme of the walk, have short sentences of less than 20 words each, use simple verbs and the active voice whenever possible, present familiar language without using technical terms, use visual aids to illustrate the message and, whenever possible, stimulate audience participation with questions or suggestions for quick searches or activities (MCINTOSH, 1982; HAM, 1992 apud VASCONCELLOS, 1998, p.19).

The panels will be made up of images, maps and written material, using language that is accessible to all visitors, taking into account the different levels of education presented by the visitor profile. The information will also be displayed in Braille. At the stopping points where there will be interpretive signs, there will be two knots in the rope serving as a warning that there is such a stop.

When designing the panels, care was taken with the visual communication part, taking into account color, shapes, illustrations, angles, fonts to form a harmonious whole in the end. The signs followed the same model as those already in place at the Unit on the Collection Trail (see photos of the Collection Trail signs).

Six stopping points were determined, including the first point, which is located at the start of the trail and has the function of indicating its existence. The title of the sign, the information it contains and the text of this information are described in Table 3.

Table 3 - Information on the Jambolao Trail Interpretive Plazas

Point	Title	Information	Interpretation text
1	Jambolao Trail	Characterization of the trail and sketch of the route with the location of the interpretive plazas.	The trail runs along the banks of the Santo Antônio stream. Length: 217 meters. Walking time: 25 minutes. Level of difficulty: easy. Suitable for people with reduced mobility and the visually impaired. Respect nature, don't litter along the way. Enjoy!
2	Meeting of the waters	Information on FEENA's water bodies: Santo Antônio stream, Ibitinga stream and Ribeirao Claro stream. Hydrographic map of FEENA.	The Santo Antônio stream increases in volume as other bodies of water flow into it along the way. At this point, the waters of the Ibitinga stream flow into the Santo Antônio, which in turn flows into the Ribeirao Claro. We use its waters for the public water supply of Rio Claro.
3	Do you know what an APP is?	Definition of APP. Indication that the area the visitor is in is considered an APP. Illustration of a riverbank and its APP.	The place where you are now is considered an APP (Permanent Protection Area), i.e. an area located on the bank of a legally protected watercourse, designed to protect the soil and forest around the river (riparian forest).
4	Who can live without water?	Information on the role of water in people's quality of life. Table with the uses of water.	Water is the lifeblood of our planet. It is an essential condition of life for every plant, animal and human being. Approximately 70% of our body is made up

			of water. We use it for: irrigation in agriculture, personal hygiene, electricity generation, industrial production, human and animal consumption.
5	What causes water pollution?	Explain some of the factors that alter water quality and availability.	Some factors that interfere with water quality: Removal of riparian forest, Chemicals used in agriculture that get into the water, Untreated sewage. These factors cause the accumulation of sediment in the river (siltation), or excess nutrients in the water (eutrophication).
6	Water pollution, until when?	Show the current scenario and probable solutions with regard to water resources.	Millions of people around the world do not have access to drinking water. Only through the rational use of water can we protect it and keep it available and of good quality for our consumption. What is your role in this fight to protect water?

Source: prepared by the author (2014).

The proposed location of the interpretive panels is shown on map 8; the intention was to relate something visible on the trail to the specific information on the panel, for example, panel 2 which contains information on FEENA's water bodies is located at the meeting of two streams: Ibitinga and Santo Antônio.

Map 8- Location of the Jambolao Trail Interpretive Plazas

Location of the Interpretive Plazas on the Jambolao Trail

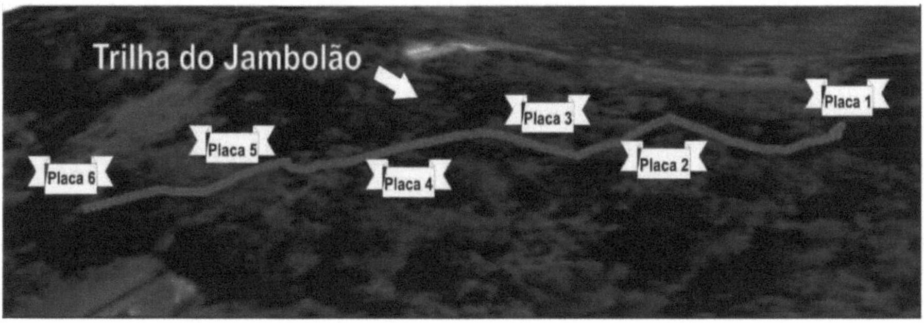

Source: prepared by the author (2014).

Following PIRES' (2001) proposal for the idealization of a system of trails at the site, it is also proposed to integrate the themes of all of FEENA's interpretive trails (taking into account the existing Collection Trail and future interpretive trails that may be created), thus disseminating FEENA's environmental, historical and cultural factors in an integrated manner. This is in line with one of the specific objectives of the Public Use Program: "To publicize the importance and characteristics of the biophysical environment, legal and historical aspects of FEENA;" (MANAGEMENT PLAN, 2005, p.177).

Along the trail, in addition to the interpretive panels, it is proposed to put up signs identifying some species of tree, containing: popular name, scientific name, origin, some curiosity or importance of the plant, what its role is for nature or for human beings.

Once the trail has been implemented, it is proposed that it be periodically maintained and assessed as to its efficiency, in order to find out whether the proposed objectives are being achieved so that it can always be improved and adapted (ANDRADE and ROCHA, 2008).

6. FINAL CONSIDERATIONS

The creation and maintenance of protected areas has proven to be an efficient measure to protect nature. In 2014 we witnessed a water supply crisis in the metropolitan region of Sao Paulo, the result of inadequate planning in the management of water resources by both the public and private spheres.

The intense deforestation of native forests, the removal of riparian forests, triggering erosion processes and silting up water bodies and springs, the contamination of rivers by the dumping of industrial effluents and domestic sewage without proper treatment, the invasion of properties on water source areas are some of the environmental factors that directly influence the lack of water in the region.

The creation and maintenance of Conservation Units, in this context, makes a decisive contribution to the quality of the air, water, soil and climate, as well as protecting species of flora and fauna, restoring the balance that existed in nature before this development model created by Western civilization of treating nature only as a commodity to be exploited and commercialized, as something outside of us that must be subjugated.

Another formidable contribution of PAs is the dissemination of environmental education, the most important measure for changing values towards a more sustainable society that respects nature. It is only by changing our attitudes and behaviors towards a more equal, just and sustainable society that we will be able to change the course of history for our species on this planet and for so many other species.

From this perspective, the planning and implementation of interpretive trails in protected areas, serving as an instrument of environmental education, is of paramount importance if we are to build a more prosperous future for everyone. The construction of a system of trails within FEENA, as well as in the other UCs, contributes to the dissemination of environmental education in the region that encompasses the unit, thus collaborating with the main objective of conserving the environmental, social and cultural aspects of the protected area.

BIBLIOGRAPHICAL REFERENCES

AGENDA 21.1992. Available at: <http://www.mma.gov.br/responsabilidade-socioambiental/agenda-21/agenda- 21-global/item/606> Accessed on: 18 Jul 2014.

ALVES, U.D.; PERINOTTO, C. C. The important historical figure who founded Fazenda Santo Antônio and led the development of the city of Rio Claro.Revista do Arquivo. Rio Claro, p. 75-78, June 2012.

ANDRADE, W. J.; ROCHA, R.F. Manual de Trilhas: um manual para gestores. Forest Institute Records Series. Sao Paulo, n 35, p. 1-74, May 2008.

BRAZIL. IBGE Census (2010). Available at: <http://cidades.ibge.gov.br/painel/painel.php?codmun=354390>. Accessed on: Aug. 14, 2014.

BRAZIL. Legislative Decree no. 2, of June 5, 1992. Convention on Biological Diversity. Available at:<http://www.planalto.gov.br/ccivil_03/decreto/1998/anexos/and2519- 98.pdf. >. Accessed on: June 22, 2014.

BRAZIL. Law No. 12.651 of May 25, 2012. Provides for the protection of native vegetation. Available at:

<http://www.planalto.gov.br/ccivil 03/ato2011 -2014/2012/lei/l12651 .htm>. Accessed on: Aug. 11, 2014.

BRAZIL. Law No. 9.985, of July 18, 2000. Establishes the National System of Conservation Units. Available at:<http://www.planalto.gov.br/ccivil_03/leis/l9985.htm>. Accessed on: Feb. 5, 2014.

BRAZIL. Law No. 9.795, of April 27, 1999. Institutes the National

Environmental Education (PNEA).Available at:<http://www.planalto.gov.br/ccivil_03/leis/l9795.htm>. Accessed on: Feb 5, 2014.

BRAZIL. Ministry of the Environment. Guidelines for Visiting Conservation Units. Brasilia: Ministry of the Environment, 2006.

BRAZIL. Ministry of the Environment. Permanent Preservation Areas and Conservation Units and Risk Areas: What does one have to do with the other? Inspection Report on the area affected by the tragedy of the rains in the Serrana region of Rio de Janeiro. Brasilia: Ministry of the Environment, 2011. 96 p.

CARVALHO, J; BOÇON, R. Planning the Route of an Interpretive Trail through Floristic Characterization.Revista Florestal.Curitiba,p. 23-32, Jan/Apr 2004.

CERATI, T. M. Proposal for the implementation of two interpretive trails for the Botanical Garden of São Paulo. 2000. 131 f. Dissertation (master's) - Center for Environmental Studies, Júlio de Mesquita Filho State University, Rio Claro, 2000.

DAHER, C.S et al.Knowing the Forest: 100 years of existence. Rio Claro, 2009.

EMBRAPA. Gleissols. Available at:

<http://www.agencia.cnptia.embrapa.br/gestor/solos_tropicais/arvore/CONTA G01_10_2212200611540.html> . Accessed on: 12 Aug 2014.

GONDIM, L. Project for an Interpretive, Sensory, Educational, Self-Guided Nature Trail for People with Reduced Mobility.

In:CONGRESO DE PLANIFICACIÓN Y MANEJO DE SENDEROS EM EL MERCOSUR, 1., 2012, Piriâpolis. Proceedings...Piriàpolis:[s.n.], 2012.Available at:<congresodesendero.wordpress.com.> Accessed on: 06 Aug. 2014.

GUILLAUMON, J. R; POLL, E.; SINGY, J.M. Analysis of Interpretation Trails. Boletim. TécnicodoInstitutoFlorestal, Sao Paulo, n.25, p.1-57, 1977.

GUIMARÂES, S.T.L. Landscapes: learning through experience. An essay on landscape interpretation and valuation. 2007. 167 f. Tese (livre docência) - Institute of Geosciences and Exact Sciences, Paulista State University, Rio Claro, 2007.

HAM, S. H. Environmental Interpretation: a practical guide for people with big ideas and small budgets. Colorado: North American Press, 1992. 456p.

IBAMA. Doces Matas Project. Manual of Introduction to Environmental Interpretation. Belo Horizonte, 2002.

LEUZINGER, M.D. Uso Pùblico em Unidades de Conservaçao. [20--].

Available at: <http://www.nima.puc-

rio.br/aprodab/artigos/uso_publico_em_unidades_de_conservacao_marcia_leu zinger. pdf> Accessed on: 07 May 2014.

LÙCIO, E. Ecological trails in the Edmundo Navarro de Andrade State Forest (FEENA): monitoring possibilities and new strategies for using space. 2011. 72 f. Dissertation (specialization in human geography) - Institute of Geosciences and Exact Sciences, Universidade Paulista Jûlio de Mesquita Filho, Rio Claro, 2011.

MARANGONI, A. M. M. C. Questionnaires and interviews: some considerations. In: VENTURI, LuisAntonio Bittar. Practicing geography: field and laboratory techniques. Sao Paulo: Oficina de Textos, 2009. p.167-174.

MENDES, A.F.; SOUZA, S.A.; TABANEZ, M.F. Interpretive Trail of the Giant Trees of Porto Ferreira State Park in Self-Guided Mode. Revista Instituto Florestal, Sao Paulo, v. 19, n. 2, p. 173-188, dec. 2007.

MENGHINI, F.B; GUERRA, A.F. S. Interpretive Trails: paths for Environmental Education. In: SEMINARIO DE PESQUISA EM EDUCAÇÂO DA REGIAO SUL, 8.,2005, Itaiai.Anais...Itajai: UNIVALI, 2005. p. 1-15.

PADUA, S.M.; TABANEZ, M.F.(org.) Educaçao Ambiental: caminhos trilhados no Brasil. Brasilia: Ipê Institute. 1997.

PAGANI, M. I. et al. Nature Trails and Ecotourism. In: LEMOS, A.I.G.(Org.).Turismo: Impactos Socioambientais. Sao Paulo: HUCITEC, 1996. p. 151-163

PIEDADE, L. I. Proposal for the Implementation of an Interpretive Trail Based on Waldorf Pedagogy in the "Edmundo Navarro de Andrade" State Forest. 2013. 64 f. Course Conclusion Paper (Graduation in Geography) - Institute of Geosciences and Exact Sciences, Paulista State University, Rio Claro, 2013.

P IRE S, A.S. Automaçao de cartas geogràficas para a identificaçao de trilhas no Horto Florestal "Edmundo Navarro de Andrade". 2001. 57 f. Course Conclusion Paper (Graduation in Geography) - Institute of Geosciences and Exact Sciences, Paulista State University, Rio Claro, 2001.

São Paulo Forestry Institute. Edmundo Navarro de Andrade State Forest Management Plan.Sào Paulo, 2005.

SÂO PAULO.Secretaria do Meio Ambiente.Manual for the construction and maintenance of trails. Sao Paulo, 2009a.

SÂO PAULO. Secretaria do Meio Ambiente. Nature Conservation Units; organizer: Luiz Roberto Numa de Oliveira; texts Adriana Neves Silva et al.Sao Paulo: SMA, 2009b. 104 p.

TRILHA da nascente. [20--?]. Available at: <http://jardimbotanico.sp.gov.br/educacao-ambiental/trilha-da-nascente/> Accessed on: July 4, 2014.

VASCONCELLOS, J. M. O. Evaluation of public visitation and the efficiency of different types of interpretive trails in Pico do Marumbi State Park and Salto Morato Nature Reserve - PR. 1998. 88 f. Thesis (Doctorate in Forestry Sciences) - Graduate Program in Forestry Engineering, Federal University of Paranà, Curitiba, 1998.

APPENDIX A - QUESTIONNAIRE APPLIED IN THE FEENA EDMUNDO NAVARRO DE ANDRADE STATE FOREST (FORMER FOREST GARDEN)

This questionnaire will help us get to know the profile of visitors to the "Horto". By answering it, you will be contributing to the proposal of a future trail for the forest. Thank you very much!

1 - **Sex:**( ) female ( ) male

2 - **Age:** ____________________________

3 -**Civil status:**

( ) single ( ) married ( ) separated ( ) widowed

4 - **What is your level of education?**

( ) Incomplete elementary school ( ) Complete elementary school

() High school incomplete () High school complete ( ) Technician

( ) Complete university degree ( ) Incomplete university degree ( ) Postgraduate

5 - **Which city do you live in**? ____________________________

6 - **What is your approximate monthly income** (based on the current minimum wage: R$ 750.00)?

( ) up to 1 MW ( ) from 1 to 3 MW ( ) from 3 to 5 MW ( ) from 5 to 10 MW ( ) more than 10 MW

7 - **How did you get to the "Horto de Rio Claro"?**

( ) car ( ) motorcycle ( ) bicycle ( ) on foot ( ) bus

8 - **How often do you visit the "Horto"?**

( )1[a] . Often ( ) up to 3 times/year ( ) 4 to 10 times/year ( ) more than 10 times/year

9 - **Who accompanies you?**

( ) come alone ( ) friends and family ( ) excursion

10 - **How long do you intend to stay in the "Horto"?**

( ) up to one hour ( ) up to two hours

( ) part-time ( ) all day

11 - **What is your MAIN activity within the "Horto"?**

( ) walk/run ( ) picnic

( ) visit to the museum ( ) other: ______________________________

12 - I would like to receive information about the "Horto" ?

( ) yes ( ) no

13 - If so:

a) **I would like to be informed about:**

( ) its waters () its vegetation ( ) its animals

() its importance () its history

( ) other ______________________

b) **How would you like to receive this information?**

( ) by employees ( ) by signs or posters (billboards)

( ) by explanatory leaflets ( ) in a visitor center

( ) in lectures ( ) on guided walks

MIX
Papier aus verantwortungsvollen Quellen
Paper from responsible sources
FSC® C105338

Printed by Books on Demand GmbH, Norderstedt / Germany